T0133354

The Language of Science

The Language of Science

From the Vernacular to the Technical

Maurice Crosland

The Lutterworth Press

The Lutterworth Press
P.O. Box 60
Cambridge
CB1 2NT

www.lutterworth.com
publishing@lutterworth.com

First Published in 2006

ISBN (10): 0 7188 3060 1
ISBN (13): 978 0 7188 3060 1

British Library Cataloguing in Publication Data
A catalogue record is available from the British Library

Copyright © Maurice Crosland, 2006

All rights reserved.
No part of this edition may be reproduced, stored in a retrieval system, or transmitted in any form or by any means, electronic, mechanical, photocopying, recording or otherwise, without the prior permission in writing from the Publisher.

Printed in the United Kingdom by
Athenaeun Press, Gateshead

To my dear wife,
who is interested in many languages.

CONTENTS

ILLUSTRATIONS

Language – a necessary tool in the
organisation of experience.

"First we have to describe [magnetism] in popular
language . . . afterwards . . . the causes of all
these [phenomena] . . .are . . .to be demonstrated
in fitting words."

William Gilbert, *De Magnete* (1600), trans. P.
Fleury Mottelay, New York, 1958, p. 26

"The terms of Art are commonly
called *Technical Words*"

John Harris, *Lexicon Technicum,* 1704

"We cannot improve the language of any
science without at the same time
improving the science itself"

Lavoisier, *Traité,* 1789

PREFACE

It is a pity if science has come to constitute almost a secret world, guarded from the general public by the use of an esoteric language. A historical approach provides an interesting and instructive introduction to understanding the situation. It is not the aim of this book to provide the reader with a key to such challenging subjects as relativity. The science of previous centuries was much simpler and easier to understand. Indeed, if one goes back far enough, much was based on little more than a common sense interpretation of the natural world, which came to be modified in the light of later experience. So the starting point of the language of science was everyday language – the vernacular. But with greater sophistication and better understanding of the natural world there was finally need for a special vocabulary – hence the sub-title of this book.

Academics regularly write learned articles and books, replete with many foot notes, perhaps to be read mainly by their professional colleagues. But the history of science is far too interesting to be confined to the university world and specialist meetings. A wider public deserves access to a variety of introductions to science. These may reveal many of the problems confronting the human spirit over the last few hundred years as well as some of the triumphs. Recently there have been hopeful signs that there is indeed a public eager to know more about the foundations of the modern world in its many aspects.

I have received several useful hints on the contents of the book from Simon Flynn and Jon Agar and I should also like to thank Daniel Pearce, the copy editor.

INTRODUCTION

In modern times one of the major obstacles, separating the general public from science is its specialised vocabulary. Indeed the different sciences all have their own specialist technical language, very useful for the professional scientist but often baffling for the lay person. Yet this has not always been the case. In the early stages of science familiar words from the common vocabulary of ordinary people were used. Only gradually was this found to be unsatisfactory. Some system had to be introduced and many new words invented to describe the growing knowledge of the natural world.

Most spoken languages have developed slowly over the centuries. Some are never written down and these often show great variations. In Shakespeare's time many words used in one part of England were not understood in other parts. Even in one place words were often spelled in different ways. Standardisation came only slowly, a feature which helps daily life but is really essential in modern science.

In the study of the natural world there were major differences at first because science as a separate activity developed only slowly. (Everything happens much more quickly today.) The word 'scientist' did not exist before the 1830s and even then took some time to be accepted. In early societies, when there were few towns, most people lived on the land and used ordinary words for the things around them, such as plants or the weather. As trade developed, some measurement of quantity was necessary and the measures introduced were often derived from the human body, such as the width of the thumb, the length of the arm or the stride of a man. This could only provide a rough measurement and might not even be recognised in a neighbouring region.

In early economies the basic problem for most of the population was sheer survival and regular hard manual work in the fields might not even guarantee this. Yet a small leisured class had time for entertainment or contemplation, and there were even a few who tried to make sense of the material world in which they lived. From the speculations of the philosophers of ancient Greece some early science emerged.

By the sixteenth century ideas had developed in the western world to the stage that not everyone accepted the common sense view that the Earth was the centre of the universe. Various plants were cultivated and a few people began to consider how they were related to each other – the basic problem of classification. Plants were often given fanciful names. Some basic chemistry had developed in various trades, such as the smelting of metals, the dyeing of cloth, yet most processes depended on trial and error rather than any theory. Chemical changes were not understood and there were still alchemists who believed that base metals like lead could, by some tortuous process, be converted into gold.

The seventeenth century is traditionally described as the period of the 'scientific revolution', when there were significant advances in physical science associated with such names as Galileo and Newton. The invention of the telescope opened up a greater knowledge of the heavens while the microscope revealed another previously unknown world. There was much work to be done in exploring these new realms, yet the workers were few, even after the foundation of the Royal Society in 1660. The following century was in some ways a period of consolidation but it was much more than this. There emerged so much new information and so many new ideas about the natural world that there needed to be some organisation of knowledge. In three particular areas, where chaos threatened, this was especially true: botany, chemistry and measurement, which are the subjects of successive chapters of this book.

Previous scholars have done research on all of these subjects but it seems that no-one has appreciated an important connection between them.

In making this claim, the author argues that the late eighteenth century has not previously been understood as such a crucial period in the history of science, a period which laid the foundations for later science in many different areas.

ONE

THE LANGUAGE OF SCIENCE IN THE EIGHTEENTH CENTURY

The need for a specific language of science

Many people who have some interest in the history of science may think of it as a succession of great discoveries. Of course no one would want to argue that that the discovery of, say, the circulation of the blood or of a new planet was not important. Yet science is more than a number of miscellaneous 'discoveries'. There are times when it is necessary to review existing knowledge in order to codify it, to examine the relationship between similar objects, to view the data from a fresh perspective or rationalise and standardise measurement. The subjects chosen here happen to be reasonably representative of the spread of eighteenth-century natural science, although the main reason for their choice in this book is that their necessary transformation was largely a linguistic one. If botany is representative of the life sciences and scales of measurement have always been fundamental in all the physical sciences, chemistry arguably fits somewhere in between in the spectrum of the sciences.

This is a book about language, but much more as well. In the first place we must of course consider the choice of names for different aspects of the natural world. But the book is basically concerned with reform, the necessary reform for the orderly development of different branches of science. The reforms we are concerned with were not minor internal changes but the comprehensive remodelling of a whole science. To undertake such reform required great ambition. For example, no one doubts the ambition of Antoine Laurent Lavoisier (1743-94), who spoke of bringing about a revolution in chemistry. Even greater was the political and social ambition of the French Revolution of 1789.

That revolution was intended to create a new world and, as a potent symbol of the fresh start, the revolutionaries abandoned the Christian calendar in favour of a political calendar, in which the foundation of the Republic in September 1792 was considered the beginning of modern history. Other countries felt no need for such a calendar and after a decade it was abandoned. This temporary political legislation contrasts with the permanence of the scientific reforms introduced in botany and chemistry. The metric system too was to achieve a permanence well beyond its country of origin.

Also there is the important question of authority. What right did a private individual in the eighteenth century have to tell his colleagues, not only in his own country but throughout the civilised world, that they should abandon the language they had always used in favour of a whole new system? This would involve men of science, many of mature years, not only learning a whole new vocabulary but also perhaps being forced to look at their subject from a new perspective. That is why the final chapter is devoted to international agreement on nomenclature.

The language of science is here really only a language in a very limited sense since it is essentially a written language. We are hardly concerned with grammar or the construction of sentences. As a collection of artificial languages, they have been constructed to avoid ambiguity. Although at times of change some scientists complained that they were being asked to learn a new language, in the case of many sciences including chemistry, it was really little more than a new vocabulary – a study which in linguistics is called lexicography, together with the occasional use of different suffixes (morphology). In botany, new but simple prescriptive rules were introduced. The strangest aspect for us today is that it perpetuated the use of a dead language. In the eighteenth century resistance was minimal because Latin was still a recognised main subject in the education of the middle and upper classes.

A connected series of developments

This book focuses on the organisation of science. But the word organisation is ambiguous. It could refer to scientific societies – a different but no less important aspect of scientific advance. Here the term organisation is used to refer to the establishment of systematic names, helping to bring order out of chaos. We begin with Carl Linné (1707-78), usually known by his latinised name as Linnaeus. He is recognised as a major figure in the history of botany, yet he was hardly a great original thinker. Rather he was a prodigious worker with an exceptional memory capable of keeping in his mind details of hundreds of plants at a time, a great codifier and synthesiser.

We then pass on to chemistry, which had a little-known connection with botany. The focus is on Lavoisier and his colleagues. Lavoisier cannot claim to have discovered any new substance, although he gave a new name and a new interpretation to the gas he called oxygen , which others had prepared earlier. His genius was to build up a whole new chemistry around oxygen and to provide a basic list of simple substances. Complementing the new theory was a new series of names which bound together the new chemistry. If one was prepared to speak of 'oxides' and 'sulphuric acid', one was more than half way to accepting the oxygen theory.

In the chapter on the metric system, we witness the introduction of standard measures in a decimal language. French scientists were organised to undertake a number of tasks to establish the data on which the system was built. No one discovered the metre. Rather it was invented as a convenient fundamental unit, the basis of a new language.

The respective subjects of the four following chapters have all previously been studied separately but they have never been connected, a task to be tackled here. The new nomenclature of plants, basically that still used today, was introduced by Linnaeus in 1753.

It was a fellow Swede, his former student Torbern Bergman, who was inspired by Linnaeus' system to apply a similar system to chemical substances, notably salts, the fundamental compounds in the chemistry of the time. But the names introduced by Linnaeus and Bergman were in Latin, a dying language. Accordingly the Frenchman Guyton de Morveau translated these names into his own language and attracted the attention of Lavoisier. The latter was completing his own new system of chemistry at the time and was interested in forming a new chemical language to express the new ideas. The two chemists collaborated in forming a new language for inorganic chemistry, which is the basis of the terms used today. (Organic chemistry really belongs to the next century).

Passing on to the metric system, we have not lost sight of Lavoisier. He was a keen advocate of the standardisation of measures and of decimalisation. He was responsible for the first definition of the gram in terms of the weight of a cubic centimetre of water. As the treasurer of the Academy of Sciences he also had responsibility for the finances of the scientific expeditions and experiments required to establish the metre.

To recapitulate on the connections between the reforms: First the comprehensive binomial (two name) nomenclature used by Linnaeus in botany served as a model for naming salts, the central chemical compounds of mineral chemistry, in the work of Bergman and Guyton, finally organised by Lavoisier. Lavoisier, the great reformer, did not confine his efforts to establishing a new chemistry. His concern for accurate weighing in chemistry was applied to the metric system, in which he took a prominent part until his arrest and execution on a trumped up charge. Some may even perceive a possible connection between the system of suffixes introduced by Lavoisier in the naming of salts and acids, and the prefixes used in the metric system. Finally the establishment of the metric system created a precedent by calling what was arguably the first international scientific conference; such meetings became the norm in the nineteenth century for agreement on the vocabulary to be used in the different branches of science, thus uniting the human race in the study of the natural world.

The idea of a perfect universal language had been advanced by John Wilkins in his utopian *Essay towards . . . a Philosophical Language*, published in 1668 with the support of the Royal Society. Earlier philosophers had spoken of finding a language which would unite mankind and Latin was an obvious candidate. Yet by the seventeenth century Latin was beginning to give way to the various vernaculars, especially in scientific writing. Also Wilkins was going further, since he sought a language which could express all knowledge in a methodical, rational and ordered fashion that would mirror the fabric of nature. The *Essay* was discussed but had little influence except on Wilkins' friend John Ray, who was encouraged to work on a botanical classification.

In the Enlightenment movement of the eighteenth century Nature was regarded as the master. While Linnaeus was involved deeply in the study of the natural world, Lavoisier claimed that his new chemical names were in conformity with Nature. Finally the justification of the French revolutionaries for choosing the metre as the basis of their new system was that it was a convenient fraction of the distance between the North pole and the Equator and, therefore, a unit based on Nature.

The eighteenth century

The final connection between the three reforms relates to time and place. They all took place in the second half of the eighteenth century and in (north) western Europe. One might claim a special place for France, leaving Linnaeus, the Swede, as an outsider. But, despite the land of his birth, Linnaeus by his travels, supporters and influence, had soon become a pan European figure.

Historians of science used to give special attention to the seventeenth century as the period of '*the* scientific revolution'. No-one would want to belittle the achievements of Galileo, Newton and others. Yet this was not the whole of science. The fact that the 'chemical revolution' occurred only in the late eighteenth century used to be regarded as an anomaly. But one of the arguments of this book is that there were major developments, which some might even wish to describe as 'revolutions', in many periods. The foundation of the new chemistry is shown here to be not an isolated 'revolution', but part of the general (re)organisation of knowledge of the period, which is an essential part of the growth of science.

Although this book is focused on the 1700s, it is very pointedly about the *second half* of that century, whose achievements overshadowed most events in the earlier century. Even the Scottish Enlightenment fits well into this time frame. In the case of science one might say, if hindsight were permitted, that the earlier period was partly one of preparation – that what happened after the mid-century built substantially on what had gone before. The respective histories of botany and chemistry provide some evidence of this. Also the earlier period was perhaps partly paralysed by its proximity to the work of Isaac Newton (d. 1727), with its overpowering influence on science generally. Newton's two great works, the *Principia* (1687) and the *Opticks* (1704) really belong to the seventeenth century.

In making the distinction between the early and the late 1700s, there are several interesting parallels. The traditional date of the beginning of the 'Industrial Revolution' in Britain is 1760. Given that this was near the period of the American Revolution (1775-83) and the French Revolution of 1789, one might conclude with the poet Wordsworth that the late eighteenth century was indeed a great time to be alive. An alternative perspective, that of the historian, is that there is some exceptionally interesting material to study.

The language of science

The analysis of the period chosen here focuses on language. Although often overlooked, language is an important part of science. Scientific language has to eschew personal feelings and fancies, and aspire to objectivity and universality. Furthermore, scientific terms must adhere to a general system with clear rules to replace the piecemeal terminology of previous ages.

To take the negative first. Clear thought can be obstructed by misleading names. In chemistry the similar names 'oil of vitriol' and 'oil of tartar' (names based on their appearance) were unhelpfully given to two chemical opposites: a concentrated acid and a concentrated solution of an alkali. In measurement the unit of length called an *aune* in France could vary in different parts of the country by more than one hundred per cent. In chemistry a positive move would be to give similar names to similar substances and in measurement to insist on standardisation, both being basic improvements introduced by the eighteenth-century reformers. By saying that plants should be known by a simple double name, giving the genus and the species respectively, instead of by long descriptive phrases, Linnaeus not only improved intelligibility but enabled easy comparisons to be made.

Language can, therefore, be used as an analytical device, as was pointed out by the abbé Condillac in 1780, a source of inspiration for Lavoisier. Because language can easily become the basis for prejudice,

Condillac urged that in our thinking we should try to forget what we have learned and go back to the beginning. Lavoisier was encouraged to think of chemical reactions on a parallel with algebraic equations (see chapter three).

Since the nineteenth century it has been customary to organise international conferences to agree on matters of scientific nomenclature. Only a general gathering of scientists within a speciality from a wide range of different countries would seem to have the authority to introduce new terminology, let alone introduce a complete new system of names. In the eighteenth century only the most distinguished representatives of a particular field would have any hope of introducing a reform of nomenclature, and even then they would be wise to bring in several colleagues to provide initial support.

In the case of botany the task fell to Linnaeus, a rising star in botany from Sweden, who had travelled extensively in northern Europe. In chemistry there was also a feeling that the language was in need of reform. A move in 1782 by Guyton de Morveau was a step in the right direction, but it was the new chemical theory of Lavoisier which justified a more basic reform. Lavoisier was wise enough to associate his manifesto of 1787, not only with Guyton, recently converted to the new chemistry, but also two other leading Parisian chemists.

For the metric system the greatest stimulus came from the strong desire in the 1780s and 1790s for political and social reform in France. This time scientists were not the initiators of change but the agents of government. They were asked to undertake the necessary research upon which the new system could be based. It was because the French Revolution was so wide-ranging in all aspects of the reform of society that weights and measures were included in the general program of reform.

There were several occasions in the nineteenth century when in Britain and the U.S.A. there was the feeling that these countries too should adopt the metric system. In 1821 the American John Quincy Adams was asked to report on whether the U.S.A. should join in the metric system. He was very enthusiastic in support and spoke of his hope that sometime in the future 'one language of weights and measures will be spoken from equator to the pole', a hope still to be realised.

The actual choice of new names was not very important in itself, although it was preferable to have fairly short names with some logical validity or some other means of helping the memory by association. Thus it would help if a plant had a name relating to, though not necessarily describing, its appearance. The really important thing was to have a coherent system, and people would be more ready to accept a new system if it was simple. Linnaeus had a system which was simple, but it has to be admitted that many of his names for species were arbitrary.

In the case of chemistry there was not a wide choice for the names of compounds since Lavoisier laid it down that such names must be based on their constituents. Once the new nomenclature was accepted, its logicality and simplicity were illustrated by the fact that students could learn the rudiments of chemistry much more quickly. In the introduction to his treatise on chemistry Lavoisier pointed out that previous trivial names had imposed a strain on the memory of students. It had, therefore, taken several years of constant application to get to grips with the subject.

Finally in the metric system the first names chosen for the units (e.g. *gave* for gram) were unpopular and were soon abandoned. Also the scientists had chosen a pretentious series of prefixes to denote differences by a magnitude of ten in any unit. These too were soon abandoned or, at least, revised.

From the vernacular to the technical

Before discussing the contrast between ordinary and technical terms it might be useful to begin by looking briefly at the unsuitability of ordinary language for scientific purposes. Although several of these points may appear superfluous, some are reflected in the popular subjective names given to plants, which would seem to justify fully their replacement by more scientific names.

An obvious first example is emotive language, which is far from objective. Death in many cases is a terrible experience, but it may become worse if, when some-one has been killed, this is described as murder. A consensual sexual encounter changes completely if it is described as rape. By way of contrast we may take the common use of euphemisms. It often happens that an elite unit of soldiers prefers not to speak of killing. The enemy is simply, and by implication surgically, 'taken out', thus absolving the perpetrators of any moral responsibility. When a family relative dies, those close to them may prefer to avoid using the objective word 'death' and say that the deceased has 'passed away'. Again, some people prefer to avoid the term 'lavatory' and speak, for example, of 'the bathroom', which would be a more appropriate term for a simple wash room. In America some men speak of 'the John', while in Britain the use of the term 'loo' has become more common. It is certainly inoffensive and may even be considered a *Deckname* or hidden name, such as used to be employed by the alchemists for secret ingredients, except in this case there is really no secret.

Although moral judgements are important in ordinary life, they have little formal place in the technical language of science. An example of an unhelpful pejorative word very relevant to this book is the word 'weed'. This can be either a plant judged harmful or unsightly by the gardener or it could even be a favoured plant in the wrong place, such as the middle of a well-kept lawn. (In French the categorisation is even worse, since to speak of a *mauvaise herbe* is to make a more explicit value judgement.)

A further fairly obvious example of language quite unsuitable for scientific purposes is the common practice of exaggeration in ordinary speech. It is probably pardonable for a person continually bothered by a swarm of flies to complain of 'thousands' of flies, even if an objective count might reveal that there were hardly more than a dozen repeatedly bothering the victim. When it comes to time, the exaggeration is in the opposite direction. Thus the customer in a café might be told to 'wait a second' for service or even 'two ticks' (of a clock), when what is at issue is a wait of many minutes. The absent shopkeeper who writes 'Back in 5 minutes' on his door may well be away considerably longer. Such reckless time keeping would clearly undermine scientific reporting.

As a further example of eminently non-scientific language, we may consider the introduction of the imagination. At first sight it might seem almost shocking to suggest banning the imagination from science. It is as if we were trying to represent science as a plodding activity with no place for originality. The whole history of science makes nonsense of this idea, yet we find Lavoisier in the introduction to his *Traité* (1789) warning of the danger of being misled by the imagination:

> Imagination . . .which is ever wandering beyond the bounds of truth, joined to self-love and that self-confidence we are so apt to indulge, prompt us to draw conclusions which are not immediately derived from facts; so that we become in some measure interested in deceiving ourselves.

In science there have been several cases of researchers

determined to prove at all costs the correctness of their pet theory by the imaginative use of selective evidence. It would not be appropriate for a scientist to be swayed unduly by the beauty of a flower or the memories brought back by a particular tree. As human beings we may appreciate what Wordsworth had to say about daffodils, but the language of the poet is necessarily very different from that of the scientist. James Keir, the author of a chemical dictionary, writing to Joseph Priestley in 1789, demanded that the French chemists should 'relate the facts in plain prose' rather than 'introduce the poetry of their nomenclature'. But it was only for their critics that the French technical terms were 'poetry', a term used in a derogatory way to indicate an unsuitable language for the subject.

Ordinary language is greatly enriched by the use of metaphors but in the cold world of science a metaphor could be mistaken for factual reporting. Only in the popularisation of science might metaphors prove a useful tool to illuminate an issue. Thomas Sprat in his *History of the Royal Society* (1667) argued against the use of high flown language and urged the use of plain language. Of course there is always the very real possibility that in scientific writing the author, while claiming to be dealing only with plain facts, is in fact employing a subtle rhetoric of persuasion.

We must now turn to the movement from the familiar and homely names of everyday speech towards more abstract and technical terms. It was partly a move from varied and subjective names to the standard and objective. Priestley, who made a major contribution to the chemistry of gases, began his account of experiments by using everyday words to describe their properties. He reported on the 'goodness' of different 'airs' to see if they were 'sweet' and 'wholesome'. Much of this work depended on the presence or absence of oxygen, but to say this is to jump from familiar everyday language to the later technical.

For many people such new artificial terms, whether, say, kilogram or carbon tetrachloride (now officially tetrachloromethane) are alien, even incomprehensible. Ordinary language, related to human experience, was supported by long tradition, as were most aspects of the life of ordinary people. One might claim that this ordinary language was more intuitive, based on common experience, such as the colour of a plant or mineral substance. It has been argued by Lewis Wolpert that much of science is necessarily counter-intuitive. This claim may be illustrated by taking a class of children, ignorant of chemistry, and (for greater effect) burning a piece of magnesium ribbon before their eyes. It burns with a bright flame with a little smoke rising towards the ceiling. If one then asks the class to decide whether they think something has been given off or taken in during the combustion, the former choice tends to be unanimous. It was Lavoisier who showed that, contrary to the current theory, the essential process in combustion is not something being given off but rather something invisible (oxygen) being taken in. The senses can easily be deceived. Therefore, Lavoisier had to use an alien instrument, the chemical balance, to show that, when a piece of tin, for example, was strongly heated, despite any possible smoke from the fire, the tin turned into a substance which weighed more than the original metal. Something had, therefore, been added to the metal.

Of course, with the advance of science, many have lamented the world they had lost. The poet Keats posed the question:

> Do not all charms fly
> At the mere touch of cold philosophy?

where 'philosophy', of course, meant science with its severely analytic method.

Modern advertisers have found that the strategic use of technical terms will help to sell their products. One recent example to market a skin cream boasted that it contained "amino-peptide complex", thus hoping that the mystique of such a phrase would help sales. Yet this use must be an exception. Thus, although, for the chemist, table salt has become 'sodium chloride', it would be quite inappropriate for a dinner guest to ask his neighbour to pass the sodium chloride. Even a professional chemist can keep the old names for ordinary life. In the same way we can admire the purple flowers in a neighbour's garden by saying, perhaps, "What a lovely group of foxgloves!". To use the botanical name, *Digitalis purpurea* would only suggest showing off. Yet this botanical example is a little different from the chemical one above, since the use of Latin terms has extended beyond botanists and professional gardeners to keen amateurs. Given the wide diversity of flowers, it has proved an invaluable means of identifying a particular species for purchase. The gardener can, therefore, if he wishes, enjoy the best of both worlds. Passing on to the subject of weights and measures, do we not see frequently in Britain, at the time of writing, the weights of goods on a market stall given both in grams and in pounds? It is as if the battles of yesterday have ended in a draw, although legislation will always favour one alternative.

In many ways the transition from the familiar to the technical runs parallel to the change from the anthropomorphic to the impersonal. This is most clearly seen in the metric system, where many of the traditional measures were related to the size of a human being, parts of the human body or human labour. The French revolutionaries were able to boast that they had advanced from a primitive basis of measurement to one related impersonally to the size of the earth. In chemistry many names had depended on the colour, taste, or even smell of a substance, that is to say on the immediate effect on the human senses. The new names introduced by the chemists in the 1780s were much more objective – one might say 'scientific'. The names of compounds were now going to be based simply on their chemical composition, with no place left for the imagination.

When we turn to botany we find a whole raft of fanciful names handed down by tradition. It would, however, be too simplistic to claim that Linnaeus abandoned all traditional names for a completely

Measurement based on the human body, the perch made up of sixteen feet. (Frankfurt, 1593)

impersonal ('scientific') system. He kept some continuity with many traditional names but insisted that the new scientific name should be in Latin and consist of only two words. The insistence on a binomial nomenclature in Latin qualifies this as a technical advance. There were many who would have preferred the vernacular to the Latin but botanists appreciated the advantage. John Berkenhout in 1789 went so far as to say that those who did not wish to learn Latin had no business to study botany!

It might be tempting for some people to think of Linnaeus as a hero of modern science. Yet, if we look more closely at his book *Species Plantarum,* we find that he spoke of the 'marriage' of plants in his sexual system. Such a metaphor is a reminder that the author belonged to an earlier age. Linnaeus was only one of a number of early naturalists and natural philosophers who, while making major advances, still had one foot in the old world. The Polish astronomer Copernicus (1543), for example, had spoken of the sun as “a lamp in a temple . . . surrounded by his family (*familiam)* of stars”. This leads to the conclusion that early science, as literature, felt free to make use of metaphors for rhetorical effect. With Linnaeus, ‘marriage’ was perhaps more than a metaphor, but such ambiguity was soon to disappear from the language of professional science. In the early years of the Royal Society, Thomas Sprat, reacting against the flowery language of many of his contemporaries, had criticised ‘this vicious abundance of Phrase, this trick of Metaphors, this volubility of Tongue’. Continuing his plea for brevity in scientific language, he urged the early Fellows of the Royal Society:

> to reject all amplifications, digressions, and swellings of style: to return back to the primitive purity and shortness.

Scientific writing was slowly to learn this lesson.

TWO

THE LANGUAGE OF BOTANY

The beginnings of botany

In the past people delighted in simpler pleasures, as we can see from the following quotation from an English literary journal of the mid-eighteenth century:

> Perhaps there is no study so delightful as botany. It traces nature through one of the most gay, sportive and fanciful branches of creation. It is attended with the utmost gratification to the senses. . . .

This represents, of course, only one aspect of botany. It reflects superficial pleasure, but neglects the serious business of making sense of the diversity of the vegetable kingdom and reducing it to some sort of order. The science of botany needs some classification together with a rational nomenclature. But we should start the history of botany at the beginning.

The Greek Theophrastus (4th century B.C.) is usually regarded as the founder of botany, with his two works: *Enquiry into Plants* and *The Causes of Plants*. In the former there is some discussion of classification. He divided plants into four categories: trees, shrubs, under-shrubs and herbs. Altogether he mentioned over 500 plants. In naming them, the terms he used were often borrowed from analogous parts in animals.

The study of botany was in many ways advanced by the general and ancient practice of using herbs to treat human ailments. This meant that, when universities were developed in the Renaissance period, a chair of botany would often be found in the Faculty of Medicine. The oldest professorship of botany dates from 1533 at the University of Padua. The professor of botany would naturally want to grow herbs and this led to the foundation of university botanical gardens (Padua, 1542). But botanical gardens would not be confined to universities. The *Jardin des Plantes* in Paris dates from 1635, the Chelsea Physic Garden was founded in 1673, and Kew Gardens in 1759.

There were, of course, gardens in antiquity. Indeed the story of Adam and Eve began in a garden. In European history the monastic garden holds a special place. The monks would often grow fruit and vegetables. Sometimes the apples from an orchard would be used to make cider. However, the most common feature was the herb garden. When the monasteries were destroyed under Henry VIII the monastery gardens disappeared. Thereafter the history of gardens in England continues in a secular context, often in great country houses, where display was paramount. In Italian Renaissance gardens too, like the *Villa d'Este* (1580s), design was everything, with terraces and many ornate fountains. If the gardens of Versailles were primarily an exercise in applied geometry, the English gardens of the eighteenth century saw a return to nature, with care attached to landscape and the planting of trees. In modern times domestic gardens have multiplied and gardening has become an extremely popular hobby.

Herbals

Herbals provide descriptions of plants directed towards their medicinal use. Utilitarian botany had the disadvantage from the point of view of science of concentrating on individual plants without considering how they related to other plants. Therefore, while herbals constitute an important area in the history of the study of plants, it has to be admitted that there was hardly any consideration of the classification of plants in this tradition. If herbs were arranged at all in a book, it tended to be according to the qualities that made them useful for medicine. Only in the seventeenth century did some of the herbalists take a broader view.

One of the earliest herbals was that of Dioscorides, a Greek author, whose original manuscripts have not survived. Fortunately there was a Latin translation under the title *De materia medica* (c. 77 A.D.), which contains the names and healing virtues of some 500 plants. Only summary descriptions of the plants were included but crude illustrations were added to later editions.

The invention of printing, which gave a great boost to learning generally, permitted herbals to pass on from the rare manuscript to a more affordable book. The first printed scientific book was Pliny's *Natural History* (1469). By 1478 a printed edition of Dioscorides was available. This work was regarded as a great authority even as late as the seventeenth century, by which time there were many competing authors. The tradition of accepting ancient authority uncritically was an unfortunate feature of early science, but the respect in which Dioscorides had been held in the Middle Ages had at least ensured the preservation of manuscript copies of his work. In the early years of printing there were also a few other herbals derived from classical or Arabic authors. They usually carried illustrations but the woodcuts were often no more than diagrammatic.

In the *Herbal of Apulius* (Rome, 1481) twenty-four different names were assigned to Verbena. Externally it was said to cure ulcers, swellings, and both dog and snake bites. Internally it cured liver complaints and removed kidney stones. An early English work, *The grete herball* of 1526, claimed to give the 'vertues in all manner of herbes to cure and heale all manner of sikenesses or infyrmytes'. Although medieval in character, the herbal does contain a few remedies recognisable in modern times, including liquorice for coughs, opium as a narcotic and olive oil for scalds.

The Latin herbal (1530) of the German, Otto Brunfels marked a distinct advance by virtue of its illustrations. Instead of representing plants in a conventional way, with each author copying the illustrations of his predecessors without reference to nature, the plants were represented here in a realistic way.

It could be argued that the definitive naming of plants in this context was less necessary than their identification by illustration. Yet names were still important in herbals since a patient could be poisoned rather than cured by a mistake in plant identification. In 1542 another German, Leonhardt Fuchs produced a Latin herbal, including some 400 native German plants and 100 foreign plants. It included a critical study of the plant names used by classical authors, something very useful in a time of confusion. Fuchs gave examples of errors in other herbals.

There now began a greater interest in foreign plants. The Frenchman Charles de l'Ecluse published in 1576 a description of plants he had observed on an expedition to the Iberian peninsula. Later he went to Austria and Hungary and not only described plants there but also various oriental plants that had reached Vienna from Constantinople. He was said to have added some 600 plants to the number known. Yet he was content simply to describe, without classifying, these new plants. For more distant parts of the world, from Mexico to India, important contributions were made by Spanish and Portuguese navigators.

By the seventeenth century the growing number of botanical names were in a state of great confusion with a variety of names being given to the same plant by different authors and great differences of opinion on the identification of plants corresponding to names used by classical authors. Thus what Dioscorides called 'Nasturtium' was not the ornamental flower familiar today but a kind of cress. Gaspard Bauhin's *Pinax* of 1623, containing a concordance of the names of some 6,000 species, helped clarify the situation. It was very influential, and the great Swedish botanist Linnaeus annotated his copy extensively.

Among English botanists of the Renaissance period the name of William Turner is pre-eminent. His English *Herball* was published in three instalments over the period 1551-1568. It was arranged alphabetically, the author showing no interest in exploring the relationship between plants. Turner had originally planned to write a Latin herbal, being particularly proud of his knowledge of that language and hence his claim to be a great scholar. However, he had asked the advice of physicians and they had advised him before publishing to acquaint himself better with plants in different parts of England, notably the west country. His final choice of English for his great herbal seems to have been on grounds of nationalism, partly to celebrate the plants of which England alone could boast.

One of the best known of English herbalists was John Gerard, whose massive book, *The Herball or Generall Historie of Plants* was published in 1597. Although an impressive work, there were many mistakes in this first edition. Gerard's account of the

‘Goose tree’ and the ‘Barnacle tree’, which were supposed to give birth to barnacle geese, had brought derision from his critics. (Albertus Magnus had disproved that legend as early as the thirteenth century.) The 1633 edition of the herbal, revised by Thomas Johnson, using smaller print but still running to some 1,600 pages with nearly 3,000 woodcuts, was a distinct improvement on the earlier edition and a facsimile edition has been produced for the modern reader.

*The ‘Round-rooted Crowfoot’ (*Ranunculus bulbosus*) from Gerard’s* Herbal *(1597)*

It would be natural to think of accurate illustrations of plants as a most desirable achievement. Yet excellent illustrations had the effect of discouraging verbal descriptions, since it was difficult to express in a few words every subtle detail of a plant. When Jerome Bock published the first edition (1539) of his herbal in German, he could not afford to pay for illustrations and, therefore, gave some attention to the inclusion of very full descriptions. Yet a full description, running to many lines of type, is very different from a name, and, if botanists were to be able to work together, every plant would eventually have to have a name upon which all botanists could agree.

Common names for plants

The common names for plants were perpetuated mainly in the oral tradition. The use of such names was reinforced in the popular herbals of Turner, Gerard and Culpeper, all in English. It is, therefore, worthwhile to look at some of these names, choosing mainly examples which are still in use today.

In a pre-industrial society most people lived close to the countryside and a majority made a living through agriculture. In an agricultural economy everyone was familiar with plant life, which they might use for food, fuel, medicine or decoration. Whereas animals might tend more frequently to have standard names within a language, there was much more variety in the names of plants in different parts of the country. Thus the Marsh-marigold or Kingcup had more than 80 local names in Britain (e.g. Marsh-blob, Horse-blob), about 60 in France and over 100 in German-speaking countries.

Plant names fall into several different categories. The easiest to explain and justify is the purely descriptive name. Thus we still speak today of blackberries, stinging nettles and perhaps pink sorrel. The snowdrop is an easily understandable name for a plant with white hanging flowers, often blooming through the snow. Then there are those that may not be so immediately obvious, but on reflection can be seen to allude to the colour of the flower and/or its shape, e.g. buttercup. The egg plant has fruit which swell to egg shape. The shepherd's purse has seed cases reminiscent of the pouch which shepherds might have hung from their waists. Then there is Jack-by-the-hedge, a roadside hedge colonising plant, otherwise known as Garlic mustard, from its smell when the stem is broken. Few people who eat grapefruit for breakfast probably appreciate that the name comes from the fact that grapefruit grow in bunches like huge grapes.

There are also more imaginative names, sometimes even poetic, such as: Love-in-the-mist, Traveller's joy (Gerard) and Enchanter's nightshade. If there are also occasional curious names like Gallant soldier, the explanation is that this is a simplified vernacular version of the Latin *Galinsoga*, a weed which is in no way gallant or soldierly. There are also several plants with names depending on the time of year at which flowers appear. The most common example is Michaelmas daisy, which flowers in late September (Michaelmas day is 29 September). Saint John's wort is so called because it flowers around the time of the feast day of Saint John (24 June). The name Alleluia, which is still used in some parts of Britain for wood sorrel, was originally applied because it flowers around Easter time, when Alleluia is again heard in churches after the long period of Lent.

Although the origin of the name Dandelion is far from obvious, it comes from the French *dent de lion* (lion's tooth, *Lowenzahn* in German) from the shape of the leaf. The name Canterbury bells cannot be fully explained by the bell-shaped flowers. It has been suggested that the flowers resemble the shape of the badges worn by pilgrims to the tomb of Saint Thomas of Canterbury. In Europe before the Reformation many plants were named after saints. In America the fact that many plants had names connected with the Devil or to witches, is related to a later Protestant tradition.

Several plants were named after animals. Usually this was to denote a less esteemed plant. Thus Cow parsley denotes an inferior parsley. Toad flax is a wild useless 'flax', fit only for toads. Dog rose is an inferior (wild) rose, and Horse-chestnut is a coarse chestnut. On the other hand Catmint is a name given to a plant with a strong scent and supposedly liked by cats.

In medieval times the natural world was seen as a puzzle or cryptogram full of hidden meanings available for human beings to decipher. According to the doctrine of signatures, the beneficent Creator had endowed plants with signs of their proper use by the human race. Thus herbs with yellow sap could be used to cure jaundice. Plants with flowers shaped like butterflies would cure insect bites (Porta, 1588). The apothecary and medical iconoclast Nicholas Culpeper went further in his Epistle to the reader in his celebrated *Herbal* (1652 and very many later editions):

> Hereby you may know which infinite knowledge Adam had in his innocence, that by looking upon a creation, he was able to *give it a name according to its nature*. [my italics]

This perspective attributes tremendous importance to names, particularly in so far as they were of ancient origin, which was true of a fair number. Thus the names of the rose, violet, fig, cypress, mint, hyacinth, ginger, lily and crocus are now known to go back to a very remote antiquity, even before the Greeks and Romans. Culpeper was making popular botanical names not simply relevant but fundamental to understanding their medicinal properties. Thus the diuretic dandelion was popularly known as piss-a-bed (in French *pisse-en-lit*).

In the days before 'chemical' medicines, people depended on plants as medicines and this is reflected in many old names: Kidney vetch, Liverwort, Lungwort, Purging flax and Scurvy grass, all with names which indicated the part of the body or the disease each was supposed to cure. In the pioneer period in nineteenth-century America, when self-medication was common, much trust was placed in herbal remedies and many were given medicinal names, such as: bellyache root, bone set, fever twig and cramp bark. It would mark a great advance for scientific purposes when the names given to plants related to their structure rather than to their beauty, supposed medicinal properties or religious associations.

Latin names

Whereas most of the later herbals in the vernacular would use common names, the more scholarly botanical works continued to use Latin names, although there was little agreement about their choice. What is difficult to understand is how, in botany almost alone in modern times, Latin has become accepted for the naming of plants well beyond the sphere of the professional botanist.

In the present age there has been a move away from the few other areas where Latin was still used. In dealings with the law, clients until recently were intimidated by phrases used by professionals such as: *pro bono* or *res ipsa loquitur*. In *Gullivers Travels* Jonathan Swift wrote about lawyers who had 'a peculiar Cant and Jargon of their own, that no other Mortal can understand'. In modern times solicitors in England have been instructed to use only the English translations of these phrases of venerable ancestry.

Following a tradition dating back to the Roman Empire, the Roman Catholic Mass was said in Latin until the changes introduced by the Second Vatican Council in 1965. The use of the vernacular has now been substituted, which most people would regard as a positive step. Quite recently (October 2005) there have been reports of the difficulty experienced by many (possibly younger) participants at a synod in Rome in understanding the traditional universal language. When the British holiday maker attends church abroad, whether, say, in Italy, Spain or Croatia, he or she may well have some difficulty in following the service and may miss the universal character which Latin introduced in previous ages.

The serious modern gardener, who probably was never troubled at school with the declension of a Latin noun or the recitation of *amo, amas, amat*, has had to become accustomed to this foreign language for the identification of his or her plants. For the perfectionist there is also the grammatical problem of gender. Thus in the Linnaean system of binomial names, the second (specific) name for a white flower may be *albus, alba, or album,* according to whether the preceding name of the genus is masculine, feminine or neuter. The polysyllabic Latin names of many plants may sometimes seem a mouthful, but the purchaser of a plant thus described has the satisfaction of knowing that there should be no mistake in being sold the wrong plant.

Before the time of Linnaeus confusion had been caused in commerce by the variety of names attached to the ever-increasing number of plants coming on to the market, which could lead to a seedsman being accused of fraud by a customer who used a different name. It would be very difficult indeed in the present century to introduce names in a dead language, but the Linnaean system has become so firmly established over many generations that it is now universally accepted. The international value of Latin might be best appreciated by a European or American travelling, say to China, and reading the bilingual label in a botanical garden.

Nicholas Culpeper aroused the wrath of the London College of Physicians when he published an unauthorised English translation of the *Pharmacopaeia Londinensis*, saying that its contents 'instead of being clothed in mystic garb (i.e. Latin), should be put upon a level with the plainest understanding'. There is a parallel in the sixteenth-century Reformation, when religious reformers had insisted on the importance of translating the Bible into the various European vernaculars. Indeed all three traditional professions: Church, law and medicine had been 'guilty' of using Latin, which was sometimes seen as a conspiracy to distance the respective practitioners from the common people. Lawyers might extract higher fees from clients, clerics might better claim superiority over the laity, and the medical profession might have an interest in the mystification of their knowledge and remedies.

Against such conspiratorial theories lies the fact that all three professions were the product of universities, where, up to the eighteenth century, Latin was in common use. After all the only education recognised as such before the nineteenth century was a classical education based on Latin and, sometimes, Greek.

Yet in early modern Europe Latin was gradually being replaced for many purposes by the vernacular. In the sixteenth century there was Luther's German *New Testament* and Tyndale's English translation. In the seventeenth century part of the scandal caused by Galileo was due to the fact that he published his book on the Copernican system in Italian instead of Latin, which would have reduced the impact of his writing to discussions among scholars. In France the *Discours sur la méthode* (1637) of Descartes became an early classic in the French language and French gradually became the *lingua franca* of diplomacy and of the upper classes throughout Europe, replacing medieval and renaissance Latin.

Predecessors of Linnaeus

We now move from the applied botany of the herbal to the pure botany of the flora – books concerned with the description and classification of plants without reference to their possible medical use. One of the sayings attributed to Newton was that, if he had been able to see further than others, it was because he had been standing on the shoulders of giants. Linnaeus may not have been quite the genius of Newton but he too, of course was able to benefit from the published work of his predecessors, notably the two great botanists, John Ray and Joseph Tournefort. But even before them we must mention the Bauhin brothers.

The French botanist Jean Bauhin (1541-1613) was a friend and pupil of the great German botanist Conrad Gesner. His greatest work was his *Historia plantarum universalis,* based on observations made on numerous expeditions throughout Europe. He was also the author of a book comprising a long list of plants named after saints (1591). He was able to show that often more than one species, even unrelated ones, were named after a particular saint. Such criticisms, as well as the influence of the Reformation, led to this type of nomenclature being largely superseded.

Jean's brother Gaspard (1560-1624), after a very broad education, was appointed professor of anatomy and botany at the University of Basle. His greatest contribution to botany was to nomenclature. His book, *Pinax theatri botanici* (1623), contained the names and synonyms of some 6,000 plants, but without any description. There was such confusion in names at this time that, to establish the different synonyms for the same plants, was a lifetime's work. Many of these synonyms were long and cumbersome descriptive phrases. For example, *Physalis annua ramosissima, ramis angulosis glabris, foliis dentato-serratis*, is no doubt wonderfully descriptive but difficult to write and remember. (Linnaeus was to reduce it to *Physalis angulata.)* It was said that his book, the *Pinax,* was the result of 40 years labour. In it Gaspard Bauhin avoided the customary alphabetical order, preferring to use a system of classification based on affinities. He took some steps towards distinguishing between genus and species and even made some use of a binomial nomenclature, thus having some claim to be regarded as a predecessor of Linnaeus.

John Ray (1627-1705), another predecessor of Linnaeus, was the son of a village blacksmith who, nevertheless, managed to study in Cambridge, thanks to the support of the local vicar. He devoted his whole life to the study of natural history, inspired by the belief that the study of nature revealed the workings of God in His creation. In compiling a British flora, he collected plants systematically, comparing them with plants already known. He often had difficulty reconciling his observations with the description and nomenclature used by the standard authors. For new plants he gave the fullest possible description. He travelled widely in Britain and on the continent of Europe to produce his encyclopaedic *Historia*

Plantarum, with successive editions in 1682, 1688 and 1704.

Greatly influenced by Gaspard Bauhin, he later developed his own classification, related to the principal parts of plants: the flower, calyx, seed and seed vessel, all of which lent themselves to ready observation. He discovered that seed plants fall into two broad natural divisions according to their possession of either one or two seed leaves. His distinction between Monocotyledons (one 'seed leaf') and Dicotyledons (two 'seed leaves') came to be generally accepted for the classification of flowering plants. This was a natural classification but it led to criticism by Tournefort and others. He tried to establish criteria for the basic question of the identification of species, concentrating on a small range of permanent structural characteristics. There had previously been a worrying multiplication of supposed species based on spurious differences. Ray made a special study of ferns, which he recognised as a distinct class. He was the first to speak of pollen, a term subsequently adopted by Linnaeus.

Finally we turn to Joseph Tournefort (1656-1708), Professor of botany at the *Jardin des Plantes* in Paris. It is interesting that his first book *Elémens de botanique* (1694) was translated into Latin, illustrating the fact that, in the late seventeenth century, Latin was still well known to scholars generally. Tournefort's system was an artificial one. He divided flowering plants into twenty-two classes, related to the form of the petals. It was a simple system, widely accepted in France and well known to the young Linnaeus, who was influenced by Tournefort's concentration on flower form as a basis for classification. Yet, unlike his contemporary John Ray and his successor, Linnaeus, he knew nothing about plant sexuality. Also Tournefort, like most of his contemporaries, was often guilty of using long descriptive phrases for names of plants.

Perhaps Tournefort's greatest contribution to botany was his definition of the genus (sharing basic characteristics), an identifiable grouping independent of the observer. This category was fundamental in the classification of plants. Without this the later Linnaean binomial nomenclature, based on the successive identification of the genus and species, would have been impossible. Once the genus was identified, it could receive a simple name related to its description. In addition to his contribution to the classification of plants, Tournefort had a good writing style which also contributed to making his work influential.

The botanical nomenclature of Linnaeus

By the late seventeenth and early eighteenth centuries more plants were being discovered and botanical names were becoming not only more diverse but more complex. In order to describe a species fully and accurately, long and unwieldy phrases were introduced. Of course the names used by a famous botanist, like Ray or Tournefort, would gain more adherents but there was still great diversity. There was an urgent need for standardisation, which was to be

Portrait of Linnaeus

the greatest legacy of Linnaeus.

Carl Linnaeus was born in 1707 in rural Sweden, the son of a poor Lutheran curate. He studied medicine in the universities of Lund and Uppsala but botany was always his first love. He spent the years 1735-38 abroad, visiting Germany, Denmark, Holland, France and England. In Paris he was well entertained by the brothers Antoine and Bernard Jussieu. His stay in Holland was particularly important in providing the patronage of Boerhaave and other influential people who were happy to pay for the publication of his historic *Species Plantarum*, the manuscript of which he had brought with him. Like Ray, Linnaeus was deeply religious: 'Man is made for the purpose of studying the Creator's works, that he may observe in them the evident marks of divine wisdom'. It was in this context that he expressed his belief in the fixity of species: 'Species are as many as the Supreme Being produced diverse forms at the beginning'.

In his youth Linnaeus was confined at first to a knowledge of regional Swedish flora, supplemented by knowledge gained on a journey to Lapland, but on this basis he was able to make a start with a simple classification. If he had been confronted as a young man with the richness of a tropical flora, he may never have begun to find any useful order. As he was later able to travel to other European countries he was able to take a wider variety of plants into consideration. In his *Critica Botanica*, an early work, he wrote:

> It is fated that botanists should impose wrong names, so long as the science remains an untilled field, so long as laws and rules have not been formed, on which they [can] erect, as on firm foundations, the science of botany. . . .

Linnaeus was essentially a systematiser. Completely ignoring other aspects of botany, he concentrated on the problem of classification. He divided plants into classes, orders, genera and species. The twenty-four classes he established were determined mainly by the number of stamens, while the sub-division of orders depended on the length or arrangement of the pistils. The 'sexual system' he developed was based primarily on the number, form and position of the stamens of a flower. It was admittedly an artificial system but it made a timely contribution when the growing number of plants available to European botanists had overwhelmed them.

Linnaeus' great contribution was directed to naming the genus and species respectively of plants throughout the vegetable kingdom. Although some simple binomial names had been used earlier in botany, he was the first to build up a whole system on this basis. In 1745 we find him inserting a binomial index at the end of a work in Swedish. In his subsequent work on the flora of Sweden he numbered the plants from one to over a thousand. This enabled him to refer to the plants easily by number rather than by using the usual long descriptive phrase. Yet, since the numbers were so arbitrary, he was tempted to use an easily remembered catch word to denote each species – another step in his path to the binomial system, in which the name of the genus (a group sharing basic characteristics) was followed by the name denoting the species (a subset of the genus). Thus:

Blackthorn	*Prunus spinosa*
Bramble	*Rubus fructicosa*
Cypress	*Cypressus sempervivens.*

Neither Linnaeus nor his successors were able to give a precise definition of a species but the Swedish naturalist had a genius for recognising species as opposed to varieties.

By 1749 Linnaeus had extended the use of catch words, which he now called epithets. By 1751 this had become a trivial name or specific name. Yet to make a specific name fully diagnostic, it was often necessary to add a qualifying phrase, which would have been a regressive move. On grounds of simplicity, Linnaeus therefore decided on the less ambitious plan of using only one word, even if it was no more than a trivial name. Although this no longer provided absolute description, it did serve to identify the plant. In a few cases it reflected the

CAROLI LINNÆI

S:Æ R:GIÆ M:TIS SVECIÆ ARCHIATRI; MEDIC. & BOTAN. PROFESS. UPSAL; EQUITIS AUR. DE STELLA POLARI; nec non ACAD. IMPER. MONSPEL. BEROL. TOLOS. UPSAL. STOCKH. SOC. & PARIS. CORESP.

SPECIES PLANTARUM,

EXHIBENTES

PLANTAS RITE COGNITAS,

AD

GENERA RELATAS,

CUM

DIFFERENTIIS SPECIFICIS,
NOMINIBUS TRIVIALIBUS,
SYNONYMIS SELECTIS,
LOCIS NATALIBUS,

SECUNDUM

SYSTEMA SEXUALE

DIGESTAS.

TOMUS I.

Cum Privilegio S. R. Mtis Sueciæ & S. R. Mtis Poloniæ ac Electoris Saxon.

HOLMIÆ,
IMPENSIS LAURENTII SALVII.
1753.

Title page of Species Plantarum (1753)

character of the plant, such as its colour. Otherwise he sometimes used geographical names, sometimes names derived from habitat, sometimes a name based on use in pharmacy. He also took further the earlier practice of arbitrarily using a latinised form of the name of a botanist he wished to commemorate. Occasionally the new name was simply a translation from the traditional name; thus (common) Thyme became *Thymus vulgaris*.

Linnaeus' masterpiece, the *Species Plantarum* was finally published in two volumes in 1753. In it he named about 7,700 species of flowering plants. He had the satisfaction of seeing his nomenclature universally accepted in his life time. One may well ask how, in the years before the introduction of international conferences, an obscure Swedish botanist managed to gain acceptance for his system of names without years of rivalry and controversy. Yet, although of humble origins and living on the northern periphery of European civilisation, the life of Linnaeus was anything but obscure. He had travelled extensively in northern Europe and gained powerful patrons. The publication of his *Species Plantarum* and several of his other books in Holland, well known for the quality of printing, helped enormously to gain them wider circulation. While still in his thirties, he had become President of the Royal Academy of Sciences in Sweden and was later ennobled. But even more relevant than his final social rank was his professional standing. By the 1760s he was accepted as 'prince of botanists'.

Apart from the support of Linnaeus' many former colleagues, students and other supporters, the inherent merits and genius of his work helped its general adoption. The new names may have been in Latin but they were comparatively simple and even logical. At a time when there was a need for a system which was universally acceptable, the Latin terms, far from being an obstacle, were a great advantage. His 'sexual system', as he called it, although not a natural system, was a useful tool in the organisation of the knowledge of flowers and filled a gap until such a time as a less artificial system could be formulated when more plants were known. The Linnaean system was, therefore, modified in the nineteenth century without altering the historic position of the *Species Plantarum,* which brought about the universal adoption of a binomial nomenclature.

In 1758 Linnaeus published the 10th edition of his *Systema Naturae*, the starting point of modern zoological nomenclature, with a similar binomial system. The members of the human race, placed at the summit of creation, became *Homo sapiens* as distinct from a speculative species, *Homo troglodytes,* for example. On the same principle the American grisly bear is now *Ursus horribilis.* Linnaeus died in 1778 at the age of 70.

After Linnaeus

Although it is the first edition of Linnaeus' *Species Plantarum* which has come to be regarded as the standard source, it has to be admitted that, in the midst of overwhelming detail, Linnaeus made a few mistakes, many of which he was able to correct in the second edition. The scientific botany introduced by Linnaeus was expanded by the French in Antoine Jussieu's 'natural' system of classification, published in the fateful year, 1789. Jussieu preferred arranging plants according to natural affinities rather than by sexual distinctions.

In the days before international conferences were regarded as essential in order to authenticate a universally acceptable language, the greatest authority lay in the hands of leading botanists. This had been the case with Linnaeus in the eighteenth century. In the early nineteenth century the French botanist Augustin Pyramus de Candolle (1778-1841) could claim such a position. In his *Théorie élémentaire de botanique*, published in 1813 in Napoleonic France he laid down rules for making new names, such as the priority of Latin, and the undesirability of forming hybrid words from Latin and Greek. In the case of rival claims for a new plant, he emphasised the claim of priority. Thus the name *Hypericum inodorum*, published by Philip Miller in the 8th edition of his *Gardeners Dictionary* (1768), had priority over *Hypericum elatum*, published by Aiton in 1789. It should be noted that it is not the date on which the name is first used that is relevant but the date of publication, which is easily verifiable.

Only in the second half of the nineteenth century did international conferences for each of the major sciences become the norm. At the international botanical congress held in Paris in 1867 Alphonse de Candolle (1806-93), son of A.P. de Candolle, introduced his *Lois de la nomenclature botanique*, which was accepted by the meeting. In 1905 a revised system was adopted at the botanical congress held in Vienna. It was at this same congress that formal tribute was paid to Linnaeus for providing the starting point and model for botanical names in his *Systema Plantarum*. In subsequent years minor variations were accepted at successive international meetings. From 1952 the *International Code of Botanical Nomenclature* became the accepted standard for professional botanists.

THREE

THE LANGUAGE OF CHEMISTRY

Chemistry before Lavoisier

To turn from modern botany to modern chemistry would involve a major change of subject. Yet, if we go back to the eighteenth century, botany and chemistry had much in common. For one thing, mathematics at that time had no place in chemistry. Rather, like botany, it was approached through classification. There was, for example, the traditional distinction between animal, vegetable and mineral. Chemistry for much of the 1700s was still in a natural history stage. Even in the late century Fourcroy, a colleague of Lavoisier, could publish a text book, the English translation of which had the title *Elements of Natural History and of Chemistry* (1788). Very little progress had been made except in mineral chemistry, where it was easier to begin to understand simple compounds. Metals had long been recognised as a distinct class. There was also the class of acids. Yet, as in botany, more material was being discovered every day and it was often difficult to see the wood for the trees.

Early chemistry had developed in various ways. A primitive chemical technology began in several ancient civilisations, notably Egypt. By the Middle Ages there was much practical experience in Europe in the extraction of metals, also some knowledge of alkalis as detergents, ceramics and glass, natural dyestuffs, and alcoholic liquors. Generally the practitioners of these trades were not anxious to pass on information of possible use to competitors. Alchemists, many living with their furnaces in dirt and squalor, hoped to transmute base metals into gold. Even more secretive than the artisans, when they wrote about their work they cultivated a systematic obscurity, often adopting a kind of code and not always in words. There was even the anonymous *Mutus liber*, a booklet consisting of fifteen illustrations but no text.

These illustrations drew on a rich symbolism. For example, the metals were associated with the heavenly bodies, with the Sun representing gold and the Moon silver. The figure of a powerful animal like a lion (the 'king' of animals?) might represent the very strong acid, a mixture of nitric and hydrochloric acids, *aqua regia* ('royal water'), which alone was capable of dissolving gold. In the accompanying illustration a lion is swallowing the Sun, representing a chemical reaction in which gold is dissolved in *aqua regia.* Such illustrations were very common in the alchemical treatises

Illustration representing the solution of gold in aqua regia (16th century)

of the seventeenth century. Superficially there might be thought to be some parallel between botanical illustrations and those in alchemical books. The great difference, however, was that the latter drew on the imagination to represent chemical reactions.

In the sixteenth century Paracelsus (1493-1541) directed alchemy away from making gold to making medicines. This provided a more acceptable justification for chemical practice. Paracelsus certainly gave a new impetus to chemistry but he also favoured natural magic and astrology. In his writing he preferred his vernacular German to the Latin of the universities; he also introduced fantastic new terms like *azoth* and *iliaster* to confuse his many critics. One consequence was the appearance of several chemical dictionaries to try to meet the challenge.

The chemistry (or 'chymistry') that began to emerge in the seventeenth century had a reputation not entirely favourable to its study. Many alchemists had been swindlers, since they pretended to have discovered the means of transmuting base metals into gold. Even the medical chemists were initially mistrusted with their chemical remedies of mercury and antimony, which could poison as well as cure in contrast to the safer traditional herbal remedies.

Robert Boyle (1627-91) was one of those who tried to give some rationale to early chemistry by considering matter to consist of tiny particles of different shapes and sizes. The reaction between acids and metals was explained by supposing the acids to be made up of tiny pointed particles, like daggers, which tore the metal to pieces. Boyle is remembered especially for his book *The Sceptical Chymist*, which undermined the theory of Aristotle that matter consisted of four so-called 'elements'.

Although alchemy had left chemistry with a bad reputation, it did bequeath three valuable legacies to early chemistry: the laboratory as a work place, a battery of chemical operations, and well tested apparatus. Chemistry was the first science to be practised in laboratories, tidier and more spacious than the alchemical laboratories portrayed by many artists. As heating was the most basic operation, but difficult to control, alchemists had developed a variety of furnaces, some providing a gentle heat, others a more intense heat. A water supply was desirable, and possibly a flue, to allow noxious fumes to escape.

Distillation was carried out with a still and there were flasks, beakers and crucibles. A pestle and mortar would enable suitable solids to be reduced to a powder. Apparatus for collecting gases was only developed in the eighteenth century. As for reagents, there would always be bottles of sulphuric, nitric and hydrochloric acids, to use names only introduced in the Lavoisier era. To the seven metals known in antiquity: gold, silver, copper, tin, mercury, iron and lead, it was later possible to add zinc, antimony and arsenic. By the time of Lavoisier, there were in addition: bismuth, cobalt, manganese, molybdenum, nickel, platinum and tungsten but not aluminium, nor metals like sodium and potassium, products of the early nineteenth century.

In the late seventeenth century Newton had begun to explore a displacement series for metals. Thus if an iron rod is placed in a solution of 'blue vitriol' (copper sulphate), metallic copper is deposited on the iron, a phenomenon once interpreted as an example of the transmutation of metals. A table of 'affinities' was drawn up in 1718 by E.F. Geoffroy, who summarised a whole series of chemical reactions in this way, including the iron/copper experiment mentioned above. Newton and some of his followers tried to explain chemical reactions in terms of the varying attraction of particles on the model of gravitational attraction, but this proved a fruitless approach.

In the great multi-volume French *Encyclopédie*, edited by Diderot in the mid-eighteenth century, Venel wrote the article on chemistry, remarking that it was a subject little understood by the public and certainly not a subject with a high reputation. He looked forward to a future revolution in chemistry, led by a 'new Paracelsus', someone of outstanding ability and drive. This was to be the role of Lavoisier in the next generation. It is astonishing that, in the course of half a century, chemistry was to be transformed from being a confused area inherited from alchemy to becoming a model science. One development which helped to make this possible was the understanding of the gaseous state.

In the seventeenth century Robert Boyle (despite the modern unhistorical statement of 'Boyle's law') had no concept of a gas. At that time people could only speak of 'air'. The Rev. Stephen Hales in his *Vegetable Staticks* (1727) urged his contemporaries to undertake a thorough investigation of this Aristotelian 'element'. He himself unknowingly prepared coal gas and other gases, but assumed that he had collected only air and, having measured their volume, just threw them away. It was left to the Scotsman Joseph Black in his *Experiments on Magnesia alba* (1756) to announce that he had discovered a new type of air which was different from atmospheric air. It was obtained by treating chalk or similar substances with an acid. This had the effect of releasing the 'air' that had been 'fixed' in the solid and he consequently called the gas 'fixed air', what we call carbon dioxide. It is a pity that Black and his successors continued to use the term 'air', when it would have been more helpful to introduce a term which avoided confusion with the atmosphere.

However, the spell was now broken, and Henry Cavendish prepared another gas (1766), which he called 'inflammable air' because it burned easily with a blue flame; Lavoisier was to call it hydrogen. Yet another British natural philosopher, the Rev. Joseph Priestley was to prepare a number of new gases, including the gas later called oxygen (1774). Priestley, however, gave it a name based on the then current theory of phlogiston. According to this theory, when a substance burned, or was strongly heated, it gave off 'flame stuff' or 'phlogiston'. Priestley's new gas allowed things to burn in it much more brightly than in ordinary air and Priestley, therefore, called it 'dephlogisticated air', that is, atmospheric air minus phlogiston. The theory was supported by certain evidence but it was to be overthrown in due course by Lavoisier. Priestley, however, was never convinced by the contrary evidence.

Boyle had criticised the idea of Aristotle that earth, water, fire and air were the universal principles of which all matter was composed. He was critical also of some chemical nomenclature, commenting on misleading or ambiguous names. It is interesting that his suggestion for improving on some misleading names was to use a long phrase, which is reminiscent of the state of affairs in botany before Linnaeus.

There were many minor criticisms of chemical names in the period before Lavoisier, but probably the most influential critic was Pierre-Joseph Macquer, the author of a chemical dictionary, first published in 1766. The dictionary form permitted the author to make many criticisms of the loose use of terminology.

Old names and the beginning of reform

In order to appreciate the need both for some reform of chemical names and the further need for systematisation, we require to know something of the chaos which preceded the French reform. If the following summary shows some order, this was not true for the wide variety of names in use. For example, the same substance was often known by different names. Some referred to colour: 'Spanish green' (basic copper acetate), and a name used by Priestley: 'red precipitate of mercury' (mercury oxide). The etymologies of the names *haematite* (blood-like stone), *orpiment* (golden pigment) and *verdigris* (green of Greece) all relate to their respective colours. The names 'liver of sulphur' (potassium polysulphide) and 'milk of lime' (a suspension of calcium hydroxide) are examples of indirect references to colour.

It was easier for the early chemists to characterise

substances by their physical properties. Two names in particular which were ridiculed by the reformers were 'flowers' (as in 'flowers of sulphur') and 'butter' (as in 'butter of antimony'). Yet 'flowers' was used fairly consistently to refer to fine powders obtained by sublimation, and 'butter' described the appearance and consistency, although this example was a very poisonous substance (antimony chloride). On the other hand taste was sometimes a criterion, as in 'bitter salt' (magnesium sulphate) and 'sugar of lead' (poisonous lead acetate). The alchemical association of the heavenly bodies with the metals was still current in the nomenclature of the eighteenth century. 'Mars' was used to denote compounds of iron, and silver was associated with the moon. The use of names of persons and places is reflected in such names as 'Glauber's salt' (sodium sulphate) and 'Epsom salt' (magnesium sulphate). The name 'tartar emetic' (potassium antimony tartrate) is a reminder of the medicinal applications of chemistry. There were even names like *mercurius vitae* (literally 'mercury of life') for a poisonous substance which did not even contain mercury. The least misleading names were probably long descriptive phrases (as in botany), such as: 'spirit of sal ammoniac prepared with sal alkali' (ammonium carbonate). An ideal name should contain useful information without being a descriptive catalogue.

Enough examples have probably been given to illustrate the wide diversity of names. As the number of known chemicals increased rapidly in the eighteenth century there was increasing confusion. How could anyone remember a vast number of unconnected trivial names? Criticism was common, but what was needed in the first place were names based not on obvious physical characteristics but rather on the chemical side, preferably names related to chemical composition. Thus Macquer suggested that the name 'vitriol' should be used exclusively for salts of 'vitriolic' (sulphuric) acid. In his dictionary he began to use systematic descriptions of salts, in which the name of the acid was followed by the name of the base.

The most significant step forward was taken by the Swedish mineralogist and chemist, Torbern Bergman, a former student of Linnaeus, who was strongly influenced by his use of a binomial nomenclature in botany. From 1775 he began applying a binomial nomenclature to salts, e.g. *magnesia vitriolata* (magnesium sulphate) and *argentum nitratum* (silver nitrate). Like Linnaeus, he wanted Latin to be the international language of science. Bergman's work was important enough to merit a French translation and, as luck would have it, the translator was Louis-Bernard Guyton de Morveau, who was independently interested in the reform of chemical names. For Bergman's strictly binomial *magnesia vitriolata* he provided (1780) the French *vitriol de magnésie* rather than *magnésie vitriolée*, which would have been a more exact translation. The important point was that there was now available in French a book which had almost binomial names for salts, names which represented the constituents of the compound, a major feature of Guyton's memoir of 1782 on general reform. The memoir made a major impression on Lavoisier and, at the centre of power in the Academy of Sciences in Paris, he was able to take the reform to its final stage.

Lavoisier and the chemical revolution

Even the briefest surveys of the general history of science that have been published usually have some mention of the 'chemical revolution' of the eighteenth century, but many ignore its close association with the reform of the language of chemistry. Yet the two were intimately connected. Before describing the reform of chemical terminology, therefore, we may summarise the career of Lavoisier and provide an outline of the major changes in chemical theory which he introduced.

Antoine Laurent Lavoisier was born in Paris in 1743 into a wealthy family. He had the benefit of attending a good school, the *Collège Mazarin*, which provided a mainly classical education. At university he studied law, but he was able to learn about chemistry through attending the public lectures of G.F. Rouelle. However, chemistry could be no more than a hobby, and his main occupation was as an official in a private company which collected indirect taxes on behalf of the king.

Yet Lavoisier was fascinated by science and was fortunate enough to be elected to the elite body of the Academy of Sciences at the exceptionally early age of 24. He impressed his senior colleagues with his energy and ambition and his publication of several memoirs. His wife later explained how he was able to combine his employment as a tax official with his scientific work:

> Each day Lavoisier sacrificed some hours to the new affairs for which he was responsible. But science always claimed a large part of the day. He rose at 6 o'clock in the morning and worked at science till 8, and again in the evening from 7 till 10. One whole day a week was devoted to experiments. . . .

In 1775 Lavoisier was appointed as one of the commissioners to take charge of the Royal Gunpowder and Saltpetre Administration. He was provided with accommodation in the Paris Arsenal, where he was able to convert a large room into a splendid laboratory, where he worked with collaborators and assistants.

The major theory which dominated chemistry in the third quarter of the century was the phlogiston theory. This explained combustion in terms of a hypothetical 'flame stuff' (phlogiston), which was supposed to leave a burning body. Most chemists were prepared to accept phlogiston, although many agreed that it could not be isolated. If it was a material substance having weight and it was given off when a metal was strongly heated, there should have been a loss of weight in the product. In 1772 Lavoisier, working with sulphur and phosphorus, found that the opposite was the case – there was a gain in weight. The same happened with metals. He concluded that the metal must have combined with the air or a part of the air. When Priestley visited Paris in 1774 and spoke about the new gas he had prepared, Lavoisier was encouraged to pursue the exploration of what he first called 'vital air' and later *oxigène*.

Portrait of Lavoisier

It had always been assumed that atmospheric air was a simple substance. Lavoisier was able to show that it was mainly a mixture of two gases which were soon to be called oxygen and nitrogen. He heated some mercury in an enclosed volume of air for a week. The apparatus he used was constructed to show changes in the volume of the air and this was slowly reduced, while at the same time specks of a red substance were formed on the surface of the mercury. The residual 'air' would not support a burning taper or life, and he consequently called it *azote* (without life). When the red specks (oxide of mercury) were collected and heated, 'vital air' was produced.

During Lavoisier's research he made many 'mistakes' and was led up blind alleys. Even the final naming of the gas *oxigène* was based on the erroneous idea that it was the principle of acidity. It is true that many acids contain oxygen but critics were soon to find some that did not.

Lavoisier continued his research with the help of several colleagues and assistants and, finally in 1785 he felt confident enough to deliver a full formal attack on the theory of phlogiston. There was no such thing as phlogiston. Those adhering to the theory did not even agree among themselves. It was a vague principle which was sometimes said to have weight and at other times not. It could be used to explain both colours and the absence of colour. As new experiments were performed, the properties of phlogiston were changed to meet the challenge.

In order to establish his oxygen theory and win over more people Lavoisier published a treatise: *Traité élémentaire de chimie* (1789), translated into English in 1790. Having, by his studies of water and air, overthrown Aristotle's theory of so-called 'elements', he presented his own table of elements in the modern sense of simple substances. The list included the gases oxygen, hydrogen and nitrogen but also many solids, including pure charcoal, which he named *carbone*, sulphur and the metals, thirty-three elements in all. He appreciated that more elements would be discovered in the future. It was unfortunate that among the elements Lavoisier included caloric (heat matter), which some critics interpreted as phlogiston in another form. As a further aid to the propagation of his new chemistry, Lavoisier and his colleagues founded the first permanent specialist journal of any science, the *Annales de chimie.*

The influence of Condillac on Lavoisier

Whereas Linnaeus in a pragmatic way had proceeded to reform botanical nomenclature, in the case of Lavoisier the reform was inspired by the Enlightenment philosopher, the abbé Etienne Bonnot de Condillac (1714-1780), usually known simply as Condillac. When Lavoisier decided to sum up his whole new system of chemistry in an influential textbook, he wrote a long and instructive preface, which began and concluded with long quotations from Condillac.

Condillac was a follower of the philosopher John Locke and was much concerned with the role of the senses in providing knowledge. He went on to consider the relationship between signs and ideas and concluded that it was signs that gave us control over the connection between ideas. Signs, of course, are expressed in language. Thus Lavoisier could begin his *Traité de chimie* (1789) with a few sentences from Condillac's book on logic, beginning:

> We think only through the medium of words – Languages are true analytic methods.

Indeed Lavoisier's concern for language was so fundamental that he claimed (with some exaggeration) that his only object in writing the book was to extend and explain more fully his contribution to the joint work on the reform of chemical nomenclature published two years earlier. He went on:

> The impossibility of separating the nomenclature of a science from the science itself is owing to this, that every branch of physical science must consist of three things: the series of facts that are the object of the science, the ideas that represent these facts, and the words by which these ideas are expressed.

He then went on, following Condillac, to consider how a child learns from sensations, some satisfying wants, others causing pain. This was the way of Nature, thus claiming the support of that great Enlightenment authority. A scientist learned in the same way. Interestingly, in his obsession with associating facts with words, he warned of the danger of the imagination. But "facts . . . are presented to us by Nature and cannot deceive".

As a model for the advancement of knowledge, he took the mathematician, who obtains the solution to a problem by the mere arrangement of data, reasoning in a series of simple steps. Indeed Condillac had taken algebra as an example of a useful language and Lavoisier was inspired to apply a similar model to understanding chemical reactions, by writing one of the first chemical equations. In the original French of his textbook this was written:

moût de raisin = acide carbonique + alkool

but the English translator, not realising the historic significance of this equation, provided only the following sentence:

> Since from must of grapes we procure alkohol and carbonic acid, I have undoubted right to suppose that *must consists of carbonic acid and alkohol"* [my italics].

Earlier, in his memoir 'On the dissolution of metals in acids', published in the Memoirs of the Academy of Sciences for 1782, he tried to express a reaction in more explicitly algebraic terms. Condillac's method of analysis was essentially decomposition followed by recomposition, which is how Lavoisier had tried to approach the problem. He expressed the part played by the acid, using alchemical-like symbols for its constituents: acid + water. Nitric acid was written as: nitric oxide (gas) + oxygen + water. Trying additionally to quantify the substances present, he ended up with very complex expressions but in this case he failed to unite them with the equals sign.

Equations that the modern chemist might recognise had to wait at least another generation. In 1813 the Swedish chemist Berzelius introduced modern chemical symbols, derived from the Latin for Lavoisier's elements. Thus there was Fe for iron (*ferrum*) , Pb for lead (*plumbum),* and so on. Once again a Swede was making use of a language which had not died out in Swedish universities, in order to produce symbols capable of international acceptance.

A simpler and more successful example of Lavoisier's use of this analytical method was demonstrating the composition of water. Water had always intuitively been understood as a simple substance. He showed that incredibly it was a compound of two gases: hydrogen and oxygen. In 1783 Lavoisier was informed that, when Cavendish had recently burned a stream of 'inflammable air' in the atmosphere, water was produced. Lavoisier had apparatus constructed which allowed a jet of hydrogen to burn in a stream of oxygen and found that drops of water were slowly produced. For confirmation, this synthesis had to be complemented by analysis. He heated water and allowed the steam to pass through a red hot iron tube, collecting the gas produced, which was hydrogen. Meanwhile some of the iron had been converted into an oxide of iron, proving that the steam consisted of hydrogen and oxygen.

The *Method of Chemical Nomenclature*

Whereas the oxygen theory had been almost entirely the work of Lavoisier, the associated nomenclature, representing the collaboration of four chemists, was published as a book in 1787. The collaboration arose partly because it had been the provincial chemist Guyton de Morveau, a confirmed supporter of the phlogiston theory, who had first published a paper (1782) urging the reform of chemical names on systematic lines. Fortunately he had come to Paris in the winter of 1786-87, where he had had discussions with Lavoisier and had been converted to the oxygen theory. Here then was an additional reason for reform. As we have seen in the last section, Lavoisier had deep philosophical reasons for a reform of language and, with his new theory, here was an obvious new beginning in the history of chemistry.

The four authors of the *Méthode de nomenclature chimique* were Lavoisier, Guyton, Berthollet and Fourcroy, although Guyton (only a humble correspondent of the prestigious Academy) was placed first on the title page to acknowledge his priority in urging reform. Claude-Louis Berthollet and Antoine-François Fourcroy were leading Parisian chemists, who had been won over to the new chemistry in 1785, although they had collaborated with Lavoisier for many years before then. The English translator of the *Méthode* judged that:

> The gentlemen who have produced the new names are ranked among the first chemists in Europe.

Lavoisier, as the leader of the group, wrote the first chapter in a philosophical vein, giving an overview of the reform. He explained Condillac's view of language as an analytical method. He claimed that "the science [of chemistry] cannot be brought to perfection unless the language is made true and correct". This made new chemical names absolutely essential and not just a piece of propaganda or an afterthought. He pointed out that many of the terms still used by chemists dated back to the alchemists, with the residue of distillation, for example, still called by the allegorical name of *caput mortuum* (dead head).

Lavoisier was quite explicit that the names of

compounds should consist of two names representing the genus and the species respectively. Thus for acids the word *acide* described the genus and, for example, *nitrique* might describe the species. He pointed out that most of the new names had been derived from the Greek, for example *oxygène* from the Greek for acid + generate and *hydrogène* from water + generate. It is interesting that they did not choose Latin, and it has been suggested that in France Latin was closely associated with the Roman Catholic church and Lavoisier, influenced by a major Enlightenment continental strand, was a fashionable anticlerical. (This association of Latin was quite different in post-reformation Sweden). Yet a little Latin did creep into some names.

Guyton's chapter explained the details of the new nomenclature. For example, simple substances should have a simple name and compounds should be named according to their constituents, a principle which could be readily applied to salts but would not have been practical in organic chemistry if it had been more advanced. He explained the system of suffixes to denote different oxygen contents of different acids and salts. In English translation the respective endings -*ate, -ite* and *-ide* denoted decreasing amounts of oxygen in salts. Similarly in acids a distinction was made between sulphur-*ic* acid with more oxygen than sulphur-*ous* acid. The names of metals, 'earths' and alkalis should be largely unchanged, but, as he was writing in French, he had to consider gender. For consistency all metals should be masculine, e.g., *le tungstène*, but earths and alkalis should be feminine, thus: *la silice* (silica) and *la soude* (soda).

Fourcroy's contribution was to explain further the details of the reform in a large folding table, in which the names of elements in the first column were followed by names of their compounds. Inevitably there was some overlap with Guyton's contribution. Fourcroy pointed out that changes in names had been kept to a minimum and the greatest innovation was in the system of suffixes. There followed a dictionary of old terms and new terms. The book concluded with a system of chemical symbols, devised by two assistants, and made up of little squares, circles and triangles representing the elements. However, the symbols were difficult for printers to reproduce and were quietly forgotten, as were different symbols used by Dalton in 1808. It was the system of letters based on abbreviations of Latin names for the elements which finally triumphed.

Acceptance of the new names

We must first consider how the new nomenclature was regarded in its native country. Before the French Revolution the books of academicians prior to publication had to be submitted to a committee appointed by the Academy of Sciences. Unfortunately the majority of the committee were hostile to the new theory and the nomenclature:

> It is not the matter of a day to reform and nearly obliterate a language which is already understood, already widespread and well known over the whole of Europe, and to put in

> its place a new language built on etymologies either foreign to its nature or often taken from an ancient language already unfamiliar to scientists. . . .

Nevertheless, they claimed to be neutral in the debate and agreed to the publication of the book on nomenclature. However, they said that, in the end, it would be the responsibility of the Academy of Sciences to legislate on its use. They were obviously thinking of the responsibility of the sister literary academy, the Académie Française, to legislate on the French language. Events were overtaken by the revolution, and it was the reception by the chemical community rather than any academy which would decide the fate of the new names.

An immediate criticism of the book was published by J.C. de la Métherie, editor of the leading French scientific journal. He argued that any change in names should be gradual. He described many of the names as "hard and barbarous" and he was particularly critical of the fact that the nomenclature was based on the oxygen theory, which he rejected. Among criticisms by other people was that new terms should not be based on a dead language. Nevertheless, the standing of the authors of the *Méthode* was greater than that of their critics, Lavoisier being an especially powerful figure in scientific circles. The majority of French chemists would be won over by the end of the century, helped no doubt by the further publication in 1789 of Lavoisier's *Traité*, emphasising the importance of the new language.

There were at least seven French issues of the *Méthode,* as well as translations into English, German, Italian and Spanish. There were even more translations of the *Traité*. Also parts of the book, particularly the dictionary, were published in several scientific periodicals and text books. Moreover, the authors of the *Méthode* themselves spread the use of the new terms by using them immediately in all their publications, particularly in their new periodical, the *Annales de chimie*. Guyton was especially zealous in policing the use of the new terms in the international scientific literature.

The English translation of the *Méthode* in 1788 was one of the first in any language. The translator, James St. John, had to establish the basic rules of orthography which decided that the French *f* should become *ph*; thus 'sulphuric' acid for *acide sulfurique.* (Americans have reverted to the original *f*). Secondly he thought it more conformable to the English idiom to change the French *i* into *y.* Thus *oxigène* became 'oxygene' (soon without the final *e*). The third rule, which rendered the French *sulfure* as'sulphuret', has been abandoned with time. (The foul-smelling gas sometimes leaking from school chemistry laboratories used to be called 'sulphuretted hydrogen'.) Given the huge respect that the translator had for Lavoisier and his colleagues, the French names of salts were translated as closely as possible to the original; thus *sulfate de cuivre* became 'sulphate of copper', leaving it to the nineteenth century to simplify this as 'copper sulphate', an order not possible in the French language. In this way the names of salts reverted to the strict binomial system of botany.

However, not many English people bought copies of the English translation of the *Méthode*, and the main way in which the new nomenclature came into the English language was through text books. The French had always been the leading producer of chemistry text books up to this time and, apart from Lavoisier, there were the books of Fourcroy and another leading chemist, Jean Chaptal.

It may have been noted that it was comparatively easy to translate the new French chemical terms into English. It was more difficult with German, particularly the word *oxigène*, which Lavoisier had coined from the Greek to mean acid producer. In German this beame *Sauerstoff,* where *Sauer* means acid, thus perpetuating Lavoisier's mistaken theory into modern times. In most languages fortunately the Greek derivation has been retained. It is only in German that there might be some cause for embarrassment, thus suggesting that there are some advantages in choosing dead languages to derive new scientific terms.

In the closing years of the eighteenth century several British critics, led by Priestley, deeply resented the claim to authority of Lavoisier and his colleagues, looking instead for a more democratic process in any change of language. Still defending phlogiston in the preface to a book of 1800, Priestley wrote:

> no man ought to surrender his own judgement
> to any mere *authority,* however respectable

Critics said that chemical language should be based on facts rather than theory. They rejected Lavoisier's geometrical method of reasoning, based on a few crucial experiments, preferring to consider the widest range of chemical experience. Several of Lavoisier's experiments required complex and expensive apparatus unavailable to his critics, thus undermining the normal experimental test of new work, based on confirmation by replication.

Priestley, a supporter of phlogiston to the end, complained that, to keep up with the latest French research, he had had to learn the new language. The new nomenclature was obviously theory laden, but so was much of the language of the critics, as when they spoke of 'dephlogisticated air' for oxygen. It was impossible to construct a completely neutral language. In desperation some critics even considered abandoning technical terms, thus, for example, deciding to speak of 'fixed air', instead of carbonic acid or carbon dioxide, a throwback to the 1750s.

Yet, by the early 1800s, it was more generally accepted that science had to distance itself from the use of common language and any ideal of a full democracy of language. Scientific language, unlike common language, could not evolve slowly over a long period. Chemists possessed special expertise, distinguishing them from a lay audience and justified the use of carefully constructed technical terms. Although Lavoisier's claim to represent truth and the language of Nature, cannot be accepted uncritically, in the end he was justified by the fact

FOUR

THE METRIC SYSTEM

Old Measures

Human beings had originally used their own bodies as a rule for linear measurement. Thus a yard was a man's good stride. An English ell (about 45 inches), was said to be related to the length of King Henry I's arm. Even in modern times a sailor, wishing to measure a rope, might still stretch out his arms, knowing that the distance between his hands would be very roughly a fathom, or six feet.

The legal measures listed in the English education code of 1900 and a part of English education for the first half of the twentieth century, set out the following units:

Length: mile, furlong, rod, pole or perch, chain, yard, foot, inch. One can hardly imagine an enormously long rod or pole of 5 1/2 yards, but the shorter unit, the chain, could be carried around to lay on the ground to make measurements. The foot as the unit of length refers of course to the foot of an adult male, whereas the origin of the inch can be better understood in its French equivalent (12 to the foot) the *pouce*, the thumb.

***Weight*:** ton, hundredweight, quarter, stone, pound, ounce, dram, grain. Here the term 'hundredweight' suggests its literal origins in one hundred 'pounds' and the homely 'stone' is only perplexing in its supposed representation of a standard weight (now 14 pounds). Less than a hundred years ago a stone could vary from 8 to 20 pounds in various parts of Britain.

***Capacity*:** coomb, bushel, peck, gallon, quart, pint. In a nineteenth-century defence of traditional measures it was argued that the bushel and the peck were well suited for men's backs, arms and hands. The pint still represents a standard measure of beer in British public houses.

***Area*:** This included 'acre', which originally represented the amount of land that a yoke of oxen could plough in a day ('acre' from the Latin *ager*, a field). Other measures related to the work output of human beings.

That so many of these units had anthropomorphic origins is hardly surprising. It would have been strange indeed if our ancestors had chosen abstract units with no relation to their daily life. One of the great problems, however, appreciated even in the Middle Ages, was the standardisation of some of these units. There were various instances when an iron rule representing a standard length was sealed into a wall so as to be accessible to the public. Thus in Paris in 1554 such a standard was available, representing the *aune*, the standard measure of drapers, corresponding roughly to the English yard. Yet in practice the *aune* differed in length for different cloths.

From the seventeenth century onwards concern was expressed in several European countries about the possibility of standardisation of weights and measures, since units of the same name were often interpreted variously in different towns. This was certainly the case in eighteenth-century France, where the variation was increasingly seen as a scandal, particularly offensive to the principles of the Enlightenment. But while people in Britain and France could urge reform, it was very difficult to imagine how this could be implemented. The French Revolution of 1789 was to provide the opportunity. Moreover, it would now be no longer a question of tidying up the old units but rather of introducing a completely new and integrated system. It should be a system simple and straightforward enough for everyone to understand. Condorcet argued that all citizens of the new republic should be self-sufficient in all calculations related to their interests. The peasant should no longer need to fear being cheated in the market place.

Early stages of reform

As early as the seventeenth century there had been ideas about a standard of measurement based on the length of a pendulum. Galileo had demonstrated that the period of swing of a pendulum for a small displacement depended entirely on its length. Several people had suggested that the length of a pendulum beating seconds could be used as a unit of length but the idea was not taken any further. Such a major change would require at the very least full government support.

In England in 1788 Sir John Riggs Miller had begun to study the problem of the diversity of weights and measures. In February 1790 he made a speech in the House of Commons proposing the choice of a general standard, from which all weights and measures might be taken. Such a standard should be derived from Nature so as to be "invariable and immutable".

Talleyrand, soon to become the French Minister of Foreign Affairs, was informed of this proposal and wrote to Miller in May 1790, expressing interest in collaboration between the two countries, to be represented by the Paris Academy of Sciences and the Royal Society of London. The Academy appointed a Commission of Weights and Measures to consider how to proceed.

If the length of a pendulum was to be chosen, the crucial question was the choice of place, since the gravitational attraction varies slightly with latitude. Talleyrand favoured a latitude of 45°, half way between the N. pole and the equator. This passed through France, and a location near Bordeaux at sea level was proposed. Miller, however, wanted London to be chosen, while Thomas Jefferson, another supporter of a universal standard based on the seconds pendulum, favoured a latitude based on the geography of the United States. Unfortunately, with the revolution in France, the political situation at the time was not favourable to international collaboration and in March 1791 the Academy of Sciences announced that it intended to proceed unilaterally.

Meanwhile the Academy Commission had decided against the seconds pendulum, since the unit of time, the second, depended on astronomical factors and was not obviously a natural and unchanging unit. It decided instead to adopt as the natural unit of length a measurement based on the size of the Earth. This unit, the metre, should be one ten-millionth part of the distance from the N. pole to the equator, drawn along a convenient meridian. Several previous geodesic measurements along the Paris meridian were available but it was thought that even greater accuracy could be attained if new measurements were taken over some 10 degrees of latitude, stretching from Dunkirk in the north to Barcelona in the south, both being at sea level. The ten-millionth fraction was chosen because it came very near to the length of the Paris *aune* (roughly three feet).

The Academy was encouraged by the fact that Earth measurement in France had a long history. The French were in an ideal position to think in terms of Earth measurement since, earlier in the eighteenth century, French scientists had led the world in triangulation procedures (see below) to construct splendid maps of France. Also it already had relevant data based on the Paris meridian. The Academy's proposal was accepted by the National Assembly on 26 March 1791. Britain was to follow the French example in map making when the Ordnance Survey was established also in 1791.

The Commission of Weights and Measures

The Commission was divided into several committees to work out different aspects of the new system. With the optimistic assumption that appropriate measurements could be carried out within a year, a generous budget was agreed. The most onerous task was the measurement of the meridian, which depended on triangulation measurements. These involved the establishment of a series of measured base lines and the very accurate reading of angles within a series of imaginary triangles formed by prominent features of the landscape. The astronomer Jean-Baptiste Delambre was to be responsible for the northern part of the Paris meridian, while his colleague Pierre Méchain was to survey the south.

After the collapse of central authority, civic control had passed to the towns and there was great suspicion of strangers, not only there but also in the surrounding countryside. The surveyors, therefore, had to furnish themselves with passports and full accreditation if they were to travel freely. Indeed political mistrust was to be almost as great an obstacle to the survey as bad weather. In a flat countryside church steeples provided excellent

Simple triangulation illustrated (16th century)

survey stations. Otherwise the scientists had to rig up their own observation towers.

The survey was to take many years and the morale of the survey teams was not improved by the news in August 1793 that the government had decided to go ahead with the metric system based on the provisional metre, using existing data. In other words it might seem that their work was not strictly necessary. Another message told Delambre of the suppression of the Academy, which further lowered his morale. Then in January 1794 he received a letter from the Commission of Weights and Measures, telling him that he had been purged from the Meridian Survey for political reasons, together with several colleagues including Lavoisier. He was instructed to hand in all his readings. Altogether political interference kept Delambre away from his survey for 18 months.

During all this time his colleague Méchain had more difficult territory to survey in the south. Free movement on the Spanish side of the Pyrenees in time of war was very difficult. Communication with Paris was constantly interrupted. Not only did he have to deal with mountain terrain but he found himself dangerously near to battles fought between the French and Spanish armies. He was arrested in Barcelona and was not able to resume his measurements until the summer of 1795. The greatest trial suffered by the scrupulous Méchain came in March 1794, when he discovered a discrepancy in his measurements, a problem he tried to conceal.

Meanwhile, back in Paris, the other members of the Commission were hard at work. Lavoisier and Haüy were to determine the new unit of weight, later called the gram. Special copper vessels were made by the famous instrument maker, Fortin. With the greatest possible accuracy the scientists measured the weight of a cubic decimetre of rainwater at the temperature of melting ice. These experiments were almost completed when Lavoisier, unpopular as a taxation officer, was arrested and executed in May 1794, a tremendous loss to science. He was never to appreciate that water was a curiously abnormal liquid with a maximum density at 4°C (more accurately 3.98°C.). This was only discovered after his death when Lefèvre-Gineau carried out a number of experiments with water at different temperatures. He

Contemporary illustration, showing the replacement of traditional measures by metric units.

was thus able to redefine the gram as the weight of a cubic centimetre of distilled water at its temperature of maximum density, rather than O°C.

Lavoisier had been a central figure in the Commission, as both secretary and treasurer and responsible, for example, for supplying funds to the two astronomical teams (far away from Paris), who were measuring the meridian. It was Lavoisier who reported on the progress of the work of the Commission to its political masters. It was also he who assembled some thirty skilled workers to assist in the metric project. He had to argue that they should be exempt from conscription. By 1793 he had, with difficulty, to justify to the new political assembly, the Convention, the many sophisticated measurements necessary to produce a comprehensive and unchallengeable new system of weights and measures.

There had to be one central permanent standard. The first standard metre was made of copper but, for the definitive metre, newly-discovered platinum was chosen, partly because it would not corrode. Although, on heating, its expansion was minimal, a clever device was introduced to measure even the very slight increase in length in hot weather. Platinum, however, was very difficult to work because of its high melting point. In 1792 Lavoisier constructed several porcelain ovens, fed by oxygen, in order to reach the highest possible temperature. He was successful in producing the platinum required for several standard measures.

Explaining the metric system to the public

Although there was great idealism and even moderation in the early years of the Revolution, the destruction of the Academy in August 1793 was a sign that the Jacobins then in power had little respect for learning, which helps to explain the orders to stop work mentioned above. The members of the Academy Commission had thought, as they were doing work ordered by the government, albeit an earlier government, that the Academy was safe, but that proved a false assumption. The members of the (now Temporary) Commission would have to work without the support of the Academy.

The early justification for the metric system was, above all, as a means of rationalising a chaotic practice of measurement. It would involve much more than introducing uniformity in weights and measures. The new system was, of course, politicised and described as 'republican measures'. The many copies of a book of *Instructions,* sent out by the government to educate the public in the new system, argued that it exemplified the republican virtues of equality and liberty. In a time of war it was also necessary to emphasise patriotism.

It is doubtful whether an intellectual justification in terms of the values of the Enlightenment was likely to win over the public. Yet great effort was put into the explanation that the measure of length, on which the whole system was based, was a 'natural' unit in as much as it was related to the size of the earth. Thus a full ten pages were devoted to an attempt to explain in the simplest possible terms that the unit was based on a fraction of a line drawn from pole to pole and called the meridian. This was a good argument to demonstrate that the metre was not an arbitrary measure. Yet its justification was so far from the experience of ordinary people that it might just as well have been plucked from the air.

The government's final concern was to get the metric system accepted throughout France. There should be a common language of measurement throughout the new Republic in the same way as there should be a common acceptance of the French language replacing the several provincial languages, including Breton, Alsatian and Provençal, which had reduced national unity under the old regime.

Decimalisation

An important feature of the metric system was that it was based on a decimal scale, that is to say the number 10. The French *Instructions* spent some time debating the merits of this system compared with the duodecimal system, based on the number 12, which was already widely in use. This had the distinct advantage that it was easy not only to obtain a half (6), but also a third (4) and even a quarter (3), whereas in the decimal system only a half (5) and a tenth (1) were whole numbers. Nevertheless, the decimal system had the important advantage of making calculation much easier, as would be readily acknowledged by anyone who compared modern British currency with the old money with twenty shillings to the pound and twelve pence to the shilling.

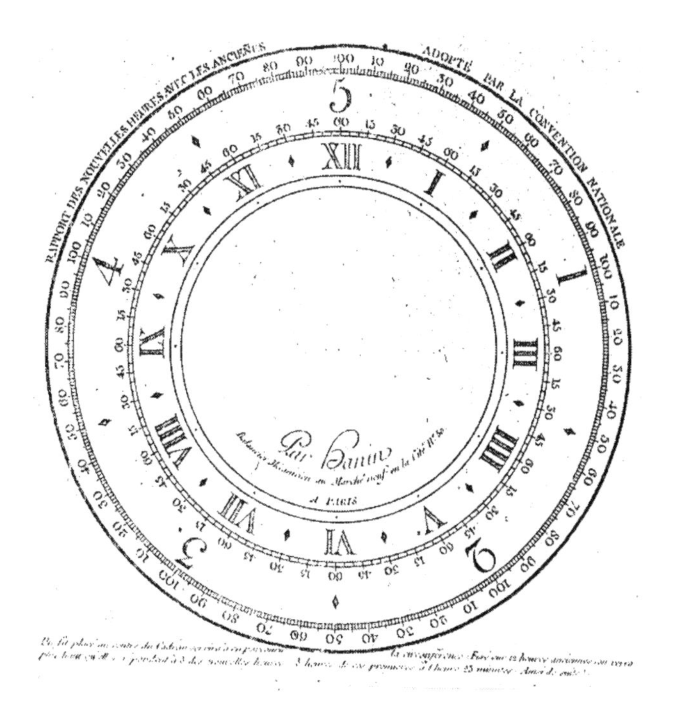

Clock face showing decimal (metric) time

Decimalisation as an example of rationalisation became so popular with the reformers that they even proposed applying it to time. The day was no longer to consist of 24 hours but was to be divided into ten (very long) hours, each hour divided into a hundred minutes and each minute into a hundred seconds. The *Instructions* pointed out triumphantly that the new republican seconds would be hardly different in length from the seconds of the old regime. The instrument maker Lenoir was persuaded to construct a clock showing the new hours and minutes.

The decimal system was further extended to

angular measure. A right angle would now have a hundred degrees. For astronomical purposes each degree would be divided into one hundred minutes and each minute of arc into one hundred seconds. One might almost speak of decimalisation fever. The idea of replacing all conventional clocks throughout France with decimal clocks was clearly impractical and was soon abandoned, while decimal angles were an easier innovation and may still be found on some French maps.

Lavoisier had long been an advocate of the decimal system. What is not generally known is that it was his intervention that persuaded the Convention to apply decimalisation to the monetary system. As a tax official Lavoisier was familiar with accounting and he argued that money should be decimalised following the lead of the metric system. No longer should commerce and industry accept the complications of a *livre* being divided into 20 *sous* and a *sous* into 12 pence, an argument that was accepted. In December 1794 the *livre* became the *franc*, divided into 100 *centimes*.

Adoption of the Metric System

On 22 June 1799 a grand ceremony was held to present the final platinum metre standards to the French legislative assemblies, who subsequently asked the French people to acquaint themselves with the new measures. There were to be tables distributed giving translations of the old measures into the new. Leaflets were circulated to introduce the system and, as mentioned above, a book entitled *Instructions* was published. Half a million metre sticks were needed for the population of Paris alone, which was to provide a precedent for the rest of the country. Unfortunately the number of sticks available did not come near to the number required and, more embarrassingly, supposed standard sticks were found to vary slightly in length.

Yet, to the dismay of the authorities and to the puzzlement of many of the scientists, neither shopkeepers nor the public were willing to give up the old measures. In November 1800 the metric system was declared to be the only system of measures acceptable for the new century, although, as a measure of compromise, the system of prefixes originally favoured, such as *deci*metre, which had frightened people, was abandoned. There was even the proposal, to allay suspicions, to return to some of the old terms for the new measures, thus *le doigt* (the finger) to represent the centimetre.

By this time Bonaparte had come to power, a figure who was ambivalent, not to say hostile, to many of the revolutionary measures. The scientists were dismayed that he refused to have anything to do with the new measures, something which seriously delayed their adoption during the Napoleonic regime, although this did not prevent some use in science (usually followed by the old measures in parenthesis).

The introduction of the metric system in France had to wait for the revolution of 1830, which deposed the reactionary Charles X and inaugurated the bourgeois monarchy of Louis-Philippe from another branch of the royal family. It was in order to suggest sympathy with the earlier revolution as much as to modernise France that in 1837 the government revived the metric system, making it obligatory throughout France and its colonies from 1 January 1840. In practice it took several generations in country districts for the new measures to be accepted but the majority of the population obeyed the new law.

In 1807 in Britain, still at war with France, John Playfair supported the metric system in principle, although with some major reservations. In Victorian Britain great pride was taken in 'imperial' measures, which ruled out any consideration of the metric system. Yet by 1857 a society was founded in support of the introduction of decimal measures internationally. In 1863 a bill to introduce the metric system in Britain and throughout the Empire was passed in the House of Commons but it failed to receive any subsequent support. It was to take another century for a major reform to be acceptable. In 1965 the British government wisely allowed a ten year period of transition, which was given a boost in 1971, when Britain agreed to join the Common Market.

There were probably two major factors responsible for delaying the introduction of the metric system in Britain. First there was the wide global extent of the British Empire (later Commonwealth), which accepted British measures. It is perhaps understandable that an early legislative step taken by India on gaining independence was one of reaction – to replace British measures by the metric system. The second factor was the persistence of non-metric units in the United States. In the early years Thomas Jefferson had supported the idea of new measures, based on the length of a seconds pendulum at some latitude to be agreed. When the alternative criterion of an Earth measurement in France was adopted, Jefferson withdrew his support.

An early value of the provisional metre by J.C. Borda was 443.5 *lignes*, where a *ligne* was a twelfth of a *pouce* (the French inch). This value was derived from the data provided in 1740 by César-François Cassini. The idea of a second determination half a century later was that a more accurate value would be obtained. It is highly ironic, when one thinks of all the hardships undergone by Delambre and especially Méchain, working in difficult terrains and in the open in all weathers, that the value they obtained for the definitive metre was 443.3 *lignes*, since with hindsight we know that this value was actually *less* accurate than that used in 1793 for the provisional metre. In any case the choice of the Paris meridian was based on convenience and the assumption of the globe as having a uniform surface, which is far from true.

In the nineteenth century doubt was expressed about the determination of the metre and the reliability of the original platinum standard, which was found to contain impurities. Yet the idea of the metre as an easily understood 'natural' unit has passed into history, although it could be argued that it had proved a useful fiction. In 1960 the metre was defined in terms of the wavelength of light emitted by a specific atom of the gas krypton. A further refinement was made in 1983, when an international conference decided that the metre should be defined as the length of the path travelled by light in a vacuum in 1/299,792,458 second. Would it be possible for scientists to depart any further from common experience?

FIVE

INTERNATIONAL SCIENTIFIC CONFERENCES

The problem of international authority

We have previously explored the historical reason for the extraordinary use of Latin in modern times for the official names of plants. Linnaeus realised that the use of his native Swedish would not be acceptable in the wider world. The most universal language in Europe in the eighteenth century was French, but Linnaeus had greater affinity with Germanic languages and, therefore, opted for the university language of Latin. This finally solved the problem of an acceptable international language.

We now come finally to the whole question of authority in science for the introduction of new names. It was by the calling of international scientific conferences that the question of authority was finally solved. By the nineteenth century it was hardly acceptable for a prince or monarch, nor even a leading scientist, to legislate on the language of science. Rather it was to the scientific community itself, first in Europe and finally in all continents, which came together to decide on a universally acceptable language for their science. Although conferences only became common in the second half of the nineteenth century, once again we may make a claim for the importance of the eighteenth century, since what may be considered as the first multi-national scientific conference took place in Paris in 1798. Moreover, it has a close connection with the previous chapter on the metric system.

First, however, we must mention the early Royal Society (founded 1660), of which Henry Oldenburg was the first Secretary. He had been keen to involve men of science from other countries by correspondence. The fact that the Royal Society and the Paris Academy of Sciences each regularly elected foreign members fostered an international perspective. Offers of government collaboration came later. For example, there was a message from the King of Spain in 1773 addressed to the Royal Society, offering to send it rare plants from central and south America if the Royal Society would agree to reciprocate from the British colonies.

The Metric Conference of 1798

It was to obtain greater international approval for the metric system that France organised an international conference in Paris in 1798-99. It was the great French mathematician Pierre-Simon Laplace who originally suggested the idea at a meeting of the First Class of the newly-founded National Institute (replacing the Academy of Sciences) on 20 January 1798. The minutes of the meeting read:

> One member suggests to the Class that it would be very useful and desirable that men of science (*savans*) sent by different governments should assist and take part in the operation remaining to determine the fundamental unity of the system of weights and measures.

Laplace suggested that the government should send invitations, and Talleyrand, as Minister of Foreign Affairs, agreed. Accordingly in June invitations were sent out to the respective governments of several allied or neutral states: the Netherlands, Denmark, Switzerland, Spain and the several Italian states. It is understandable that in a time of war Britain was not invited; nor was the United States. This may have resulted in the conference being less well known than it deserves in so far as it set a precedent.

The foreign delegates were invited for September 1798, not just as observers but as collaborators in the final stages of the establishment of the metric system, since most were prominent men of science in their own countries. Yet nearly all the work had been done by the French scientists, so the guests were often reduced to no more than making minor criticisms. After an initial delay, the eleven foreign delegates met regularly with the French scientists to examine details of the system. Finally in June 1799 the whole group reported to the National

Institute and to the two legislative councils. The new metric system now had the approval of the official representatives of French science, representatives of the foreign governments and of the French government. The foreign scientists were asked to try to persuade their respective governments to adopt the metric system, not an easy task around 1800, when even the French had great problems in persuading their public to take the system seriously. Apart from the natural reluctance to abandon traditional measures, many countries had problems translating the French names of the units. Fortunately this was never a major difficulty in the English language.

From international co-operation to conferences

A distinction must be drawn between simple international co-operation and the calling of international conferences, such as the one on the metric system, which required considerable organisation. However, it was not until the second half of the nineteenth century that a new spirit of professional unity emerged, fortified by the development of railways which made travel so much easier.

Meanwhile the Congress of Vienna of 1814-5, called after the defeat of Napoleon, helped to provide a fresh spirit of international co-operation for the nineteenth century. Up to this time there had been important differences in the maps produced in different countries, some having major inaccuracies. Obviously accurate mapping of the world's oceans for maritime safety in international trade was in everyone's interest. Already in the 1820s collaboration had taken place between the hydrographic offices of Britain, Denmark and Sweden and there were plans to extend this to France, Spain and Russia. At the first meeting of the Royal Geographical Society of London in 1831 the president remarked on the importance of hydrography and the preparation of accurate charts, with greater knowledge of tides. This encouraged some collaboration across the Atlantic. Yet it was not until 1889 that the first international maritime conference was held in Washington.

Since navigation depended so much on the compass, a more detailed knowledge of the earth's magnetism was desirable. In the 1830s Alexander von Humboldt and François Arago were establishing a series of magnetic stations stretching from Paris to China. In 1836 they asked the Royal Society to co-operate in setting up further stations and co-ordinating their observations. The adoption of a uniform plan was necessary for the advancement of knowledge of the magnetic elements of the globe. Observations should not be made randomly but at an agreed time. The agreement on such a plan went far beyond the power of individuals. Already in the 1830s support had been obtained from the respective governments of France, Prussia, Hanover, Denmark and Russia.

The Belgian Adolphe Quetelet was actively engaged in the 1830s in international collaboration in astronomy, meteorology and geophysics. It was, however, his monograph of 1835 on social statistics which gave him an international reputation and the authority to call for a multi-national congress to agree on the use of the same standards in different countries to make comparisons possible. Communications were steadily improving with the electric telegraph which transcended national boundaries. In 1865 some twenty European states met in Paris to work out an international convention on the telegraph. The interaction of scientists for practical and technological purposes provided a good precedent for the organisation of pure science on international lines. The other major precedent for the organisation of international scientific conferences came from the practice of holding international exhibitions. Many conferences were planned to coincide with industrial exhibitions in the second half of the nineteenth century.

Although many of the international conferences were concerned to agree on the universal acceptance of certain standards, the question of an agreed nomenclature was prominent at most of the meetings. The desirability that decisions on the nomenclature to be adopted in different branches of science should depend on a general consensus was expressed as follows by a group of British zoologists in 1842:

> The world is no longer a monarchy, obedient to the ordinances, however just, of an Aristotle or a Linnaeus. She has now assumed the form of a republic. . . .

This looks like a call for an international meeting, but the group, which included Charles Darwin, started from the premise that Britain was a world power and they felt that the British Association for the

Advancement of Science alone would have sufficient international authority.

The case of geology differs from that of the other sciences in that most of the concern for an international meeting came not from a European country but from the United States. American geologists had encountered strata which had no name in the old world. Different names were being given to the same thing in different countries. American geologists had settled for a temporary pragmatic nomenclature relating to place, leaving the problem of rationalisation to the future. At a meeting of the American Association for the Advancement of Science held in Buffalo in 1876 it was proposed to use the Paris Exhibition of 1878 as the occasion for an international conference of geologists. Paris was a favourite city for holding international conferences.

The Karlsruhe Congress

In order to understand the reasons for calling what was arguably the first major international conference of the nineteenth century, it is necessary to fill in a little basic background. Lavoisier is rightly remembered for his oxygen theory and his list of simple substances or elements. Yet he had refused to include in his new chemistry things which he could not experience directly in the laboratory, particularly atoms. Theories of atoms go back to the ancient Greeks and such theories were certainly discussed in the eighteenth century. Yet for Lavoisier all this was sheer speculation. It was left to the English chemist John Dalton (1766-1844) to supplement Lavoisier's elements with a soundly-based atomic theory.

In discussing the atoms of Lavoisier's elements, Dalton took the important step of giving a list of the relative weight of these elements (1808). Taking the lightest element, hydrogen, as having a weight of one and, knowing that water was a compound of hydrogen and oxygen in the ratio by weight of approximately 1:8, he gave the relative weight of the oxygen atom as 8. This was based on the simple assumption that water consisted of one atom of hydrogen combined with one atom of oxygen. Those who later could argue from the experimental fact that water was formed when two volumes of hydrogen combined with one volume of oxygen, gave oxygen the atomic weight (nowadays 'relative atomic mass') of 16 and the well-known modern formula H_2O, to use the symbols introduced by Berzelius in 1813.

Atomic weights were most useful in chemistry but we have seen already that there was lack of agreement among chemists as to their numerical value. Those who gave oxygen an atomic weight of 8 were really listing its *equivalent weight*, whereas we know with hindsight that the figure 16 corresponds to its true *atomic weight.* Problems of equivalents versus atomic weights were to dog chemistry throughout much of the nineteenth century.

Although it can be claimed that Lavoisier laid the foundations of organic chemistry with his quantitative analysis of a few vegetable substances, it was only in the nineteenth century that organic chemistry developed, becoming then the most flourishing branch of the science. The carbon atom was found to have the ability to combine endlessly with other carbon atoms and also with different combinations of hydrogen, oxygen and nitrogen, producing thousands, eventually millions, of different compounds. Some of these could be extracted from the vegetable or animal kingdom, but the most exciting aspect of organic chemistry was that many more could be prepared in the laboratory, a procedure which seemed to have no limits.

It was soon possible to attach a formula to simple substances like methane, CH_4 (marsh gas), more complex substances like sucrose (ordinary sugar), $C_6H_{22}O_{11}$, and very complex compounds with long chains of carbon atoms. Eventually the German chemist August Kekulé (1829-96) discovered that, in addition to compounds with long chains of carbon atoms, there were many substances made up of rings of carbon atoms, of which the simplest was benzene, consisting of six atoms of carbon combined with six atoms of hydrogen. Organic chemistry was developing so quickly by the mid-nineteenth century that there was great confusion with the emergence of rival theories adhered to by different schools.

It was the German chemist August Kekulé who had the idea of calling an international conference to resolve disputes between chemists. He called on his friend, Karl Weltzien, a teacher at the Technische Hochschule at Karlsruhe, to organise the conference. In 1858 Weltzien had composed a large treatise on organic chemistry, in which he attempted to classify all the organic compounds then known. He complained that his greatest difficulty was the nomenclature of compounds. Even when the formula of a compound was known, a name was often lacking. After consulting various colleagues, including Adolphe Wurtz in Paris, Weltzien decided to send a circular letter to chemists throughout Europe to discuss not only organic nomenclature but even the very basic terms used in chemistry, such as atomic weight, equivalent, and the writing of formulae.

In the end no less than 140 chemists, representing nearly every country in Europe, met in Karlsruhe in September 1860. As an explicitly international gathering, the conference could not simply use the local German language but also accepted French and English on equal terms. In his opening address Weltzien said that, if they were able to agree on a more exact formulation in chemistry, it should be possible to teach the subject in a relatively short time. The confusion existing in organic chemistry in 1860 had a parallel with the situation in mineral chemistry some eighty years earlier, before Lavoisier reorganised that subject. He too had made the point about students learning the new chemistry in a much shorter time.

A committee was established to examine a range of problems, ranging from the exact meaning to be given to the words atom and molecule (sometimes then used as synonyms) to the much more ambitious question of general nomenclature. Also it was suggested that, given the persistent confusion between atomic weights and equivalents, where the former value was often twice the latter, it would be useful if the symbol for carbon, C, was used to represent an equivalent value of 6, whereas the now more generally accepted atomic weight of 12 would be represented by C with a bar through it: almost like the modern symbol for the euro (€).

Inevitably quite a few senior chemists did not take the trouble of going to Karlsruhe. The great German chemist, Justus von Liebig, was absent but lent his support to the meeting. From France Marcellin Berthelot, a leading and influential exponent of equivalents, was absent. On the positive side we may mention the presence of several young German chemists who later became leaders in their field and were able to attend the historic Geneva Congress of 1892, called specifically to reach full agreement on the nomenclature of organic compounds, a task beyond their predecessors of 1860. The calling of a succession of chemical conferences was followed by the foundation of the International Union of Pure and Applied Chemistry and other bodies established specifically to agree on such matters as universal standards of nomenclature.

Other international conferences

We may return for a final time to the history of the metric system, which we showed was the result of the work of scientists from one nation alone, however much the French government wanted to put an international gloss on the new weights and measures.

It was the holding of international exhibitions which probably contributed most to the spread of the metric system. The Great Exhibition held in London in 1851 revived an interest in Britain in metric measures and at the 1855 Paris Exhibition the panels of judges representing different countries undertook to tell their respective governments about the advantages of a universal system of weights and measures. By this time there was the danger of a scandal with the metric standards adopted being slightly different in different countries. An international conference was therefore called in Paris and 24 countries agreed to send representatives. It was unfortunate that the date chosen for the conference was August 1870, since by this time France was at war with Prussia. The main business of the conference was postponed till 1872, with a further conference in 1875, now called *Conférence diplomatique du mètre.* Standards of accuracy had now improved considerably and there was, for example, dissatisfaction with the platinum used in the original standard as a result of the impurities that had been subsequently found. The keeping of standards was considered so important that it was decided to establish a permanent Bureau in the Paris region, acknowledging the French origins of the system.

Electricity in the second half of the nineteenth century was becoming a subject of major industrial as well as scientific importance and a very basic question which needed to be settled was the definition of units. It was vital that there should be international agreement. Once again it was the British

Association which took the initiative. In 1861 it appointed a committee to make proposals and in 1862 wrote to leading physicists in all the major European countries as well as the U.S.A. to inform them of their ideas. However, it was hardly realistic to suppose that a consensus could be achieved by correspondence. The first international conference of electricians had to wait until 1881 with the primary purpose "to agree on the universal language of electricity". The conference was held in Paris, this time with the claim that it was fitting to meet in the country of Ampère and Arago. Strong nationalistic feelings prompted the French to demand that the names of Ampère and Coulomb should be commemorated in the names of units, while the Germans were keen to commemorate Gauss and Weber. Fortunately there were enough units in electricity and magnetism for all these scientists and others to be remembered. Yet rather more important than the choice of names for units of electricity was the definition of these units. It proved particularly difficult to reach agreement on the definition of the unit of resistance, the ohm. It was not until the meeting in Chicago in 1893 that the ohm was officially defined as the resistance of a column of mercury of one square mm. cross section and 106.3 cm. in length, thus further consolidating the acceptance of the metric system in science.

By the late nineteenth century international conferences had become well established for all the main branches of science. With the emergence of new sciences it therefore became important for them

FURTHER READING

Chapter One

There are many general histories of science, but in most of these there is usually a concentration on the important events of the seventeenth century, which overshadow the eighteenth century. However, for a recent survey of eighteenth-century science see: Roy Porter, ed., *The Cambridge History of Science*, volume 4, *Eighteenth-century Science*, Cambridge, 2003. This large book includes many different perspectives without focusing on scientific language. There is also much information on eighteenth-century science in: A. Wolf, *A History of Science, Technology and Philosophy in the Eighteenth Century,* 2nd edn., London, 1952. For a discussion of language, although dealing exclusively with the seventeenth century, see: Brian Vickers, ed., *English Science, Bacon to Newton*, Cambridge, 1987.

Thomas Sprat's *History of the Royal Society* (1667) has been reprinted in facsimile (Introduction and notes by J. Cope and H.W. Jones, London, 1966). Melvyn Bragg in *The Adventure of English* (London, 2003) includes brief mentions of scientific English and deals especially with the derivation of English scientific and medical terms from the Latin and Greek.

Chapter Two

Anyone wishing to study the contribution by Linnaeus to the reform of botanical names owes a tremendous debt to the labours of William T. Stearn. See especially his many introductory essays to the facsimile reprint of Linnaeus, *Species Plantarum*, 2 vols., (1753), ed. W.T. Stearn, 2 vols., London, 1957. See also: William T. Stearn, *Stearn's Dictionary of Plant Names for Gardeners*, London, 1963 and W.T. Stearn, *Botanical Latin,* 2nd edition, Newton Abbot, 1973. See also: Geoffrey Grigson, *A Dictionary of English Plant Names*, London, 1974 and David Gledhill, *The Names of Plants,* 2nd edn., Cambridge, 1989. A good general history of botany is: A.G. Morton, *History of Botanical Science*, London, 1981. For many years the standard work on herbals has been: Agnes Arber, *Herbals, their Origin and Evolution. A Chapter in the History of Botany, 1470-1670,* Cambridge, 1953. Most recently published, with many illustrations in full colour, is: Anne Pavord, *The Naming of Names: The Search for Order in the World of Plants,* London, 2005, covering the period from ancient Greece till the time of Linnaeus.

Chapter Three

The history of chemical language from the alchemists up to 1900 is provided by: M. P. Crosland, *Historical Studies in the Language of Chemistry*, London, 1962, 2nd edition, New York, 1978. Another study of the new nomenclature and its background is: Marco Beretta, *The Enlightenment of Matter,* Canton, MA., 1993, chapters 3 and 4. For a review of British criticisms of the new nomenclature see: Jan Golinski, 'The chemical revolution and the politics of language', *The Eighteenth Century. Theory and Interpretation,* vol. 33 (1992), 238-251. Several biographies of Lavoisier were published around the commemoration of the bicentenary of his death in 1994, for example: Jean-Pierre Poirier, *Lavoisier, Chemist, Biologist, Economist,* Philadelphia, 1996. A facsimile edition of the English translation of Lavoisier's *Traité: Elements of Chemistry*, Edinburgh, 1790, may still be available in paperback (Dover Publications, New York).

Chapter Four

The most recent detailed study of the metric system is: Ken Alder, *The Measure of all Things. The Seven-Year Odyssey that transformed the World*, Little, Brown, London, 2002, which gives special attention to the triangulation measurements of Delambre and Méchain. The standard general history is: G. Bigourdain, *Le système métrique des poids et mesures*, Paris, 1901. See also: M. P. Crosland:

'Nature and measurement in eighteenth-century France' in: *Studies on Voltaire and the Eighteenth Century,* vol. 87 (1972), 277-309, reprinted in: M..P. Crosland, (ed.), *Studies in the Culture of Science in France and Britain since the Enlightenment,* Aldershot, 1995. *Science in France in the Revolutionary Era,* ed. M.P. Crosland, Cambridge, Mass.,1969. Also Josef Konvitz, *Cartography in France, 1660-1848,* Chicago, 1987.

Chapter Five

Most of the work that has been done on international science has been focused on the twentieth century by scholars whose primary interests are political or sociological. The history of earlier international conferences has attracted much less attention. However, further information may be found in: Eric Forbes, ed., *Human Implications of Scientific Advance*, Edinburgh, 1977. The relevant material is reprinted in: M.P. Crosland, *Studies in the Culture of Science in France and Britain since the Enlightenment*, Aldershot, 1995. For the first metric conference see M.P. Crosland, 'The congress on definitive metric standards, 1798-99', *Isis,* vol. 60 (1969), 226-231. See also: K. Alder, under Chapter Four above. For the Karlsruhe conference, see: Clara de Milt, 'Carl Weltzien and the Congress at Karlsruhe', *Chymia*, vol.1 (1948), 153-169

INDEX